THE WOMEN OF THE MOON

THE WOMEN
OF THE
MOON

Tales of Science, Love, Sorrow, and Courage

Daniel R. Altschuler
Fernando J. Ballesteros

OXFORD
UNIVERSITY PRESS

OXFORD
UNIVERSITY PRESS

Great Clarendon Street, Oxford, OX2 6DP,
United Kingdom

Oxford University Press is a department of the University of Oxford.
It furthers the University's objective of excellence in research, scholarship,
and education by publishing worldwide. Oxford is a registered trade mark of
Oxford University Press in the UK and in certain other countries

First Edition published in 2019
Impression: 1

Published in the United States of America by Oxford University Press
198 Madison Avenue, New York, NY 10016, United States of America

British Library Cataloguing in Publication Data
Data available

Library of Congress Control Number: 2019937067

ISBN 978–0–19–884441–9
DOI: 10.1093/oso/9780198844419.001.0001

Printed in Great Britain by
Bell & Bain Ltd., Glasgow

Links to third party websites are provided by Oxford in good faith and
for information only. Oxford disclaims any responsibility for the materials
contained in any third party website referenced in this work.

Dedicated to our women of the Earth, Celia, and Herminia.
In memory of all women who have been and are being mistreated by men

Foreword

The Moon is feminine in Spanish – in all the Romance languages, in fact. Even in English, which doesn't assign genders to inanimate objects, the Moon, like the planet Venus, is presumed female. Her sex was likely decided by her association with the ancient chaste huntresses, Artemis and Diana, and later with the Virgin Mary, as well as by the match of the Moon's monthly cycle of changing phases with the human menstrual cycle. Although the markings on the lunar surface add up to a "man in the Moon" for most casual observers, and the only people to have trod that surface were all men, still the Moon remains womanly.

In preparing their tender paean to the 28 women commemorated in officially named craters on the Moon, astronomers Daniel Altschuler and Fernando Ballesteros have illuminated the lives of heroines both familiar and obscure. I am pleased to have played a small part in helping their book, which they originally wrote in Spanish and published as *Las Mujeres de la Luna,* reach an English-speaking audience via Harriet Konishi at Oxford University Press. Right around the time the authors appealed to me for help in finding an English-language editor, I happened to meet Harriet in Columbus, Ohio, at the April 2018 convention of the American Physical Society. I have since spent time with Fernando at his home institution in Valencia, Spain, during a March 2019 conference on the subject of women in science. Daniel and I crossed paths so long ago – in 1992, at the Arecibo Observatory in Puerto Rico – that it is high time for us to meet again. Meanwhile, we are united in this volume by our mutual interest in women whose research, observations, explorations, or other accomplishments gained them a namesake on the Moon.

Dava Sobel,
author of *The Glass Universe and Galileo's Daughter*

Contents

Pretext

Art thou pale for weariness
Of climbing heaven and gazing on the earth,
Wandering companionless
Among the stars that have a different birth,
And ever changing, like a Joyless eye
That finds no object worth its constancy?

PERCY BYSSHE SHELLEY (1792–1822)

The Moon is no longer the "in" thing[1]. We see it as often as the Sun and give it little thought—we've become indifferent. What captures our attention are those exotic discoveries that make headlines: the Big Bang, black holes, dark matter, things that are mysterious, that still lie in that space where scientific study overlaps with speculation. What can the Moon possibly tell us? The last time it made headlines was when the Apollo astronauts landed.

That was about 50 years ago, when you, dear reader, possibly were not yet born. On July 20, 1969, the astronauts of NASA's Apollo 11 mission, Neil Armstrong (1930–2012) and Edwin "Buzz" Aldrin (born 1930), became the first humans to walk on the Moon after a three-day voyage. They settled—for somewhat less than a day—at a place they called Tranquility Base, (officially: Statio Tranquillitatis), at the northern edge of Mare Tranquillitatis. A third astronaut, Michael Collins (born 1930), was waiting for them in the Command/Service Module, in orbit around the Moon. The three would return safely to Earth. The moon landing was, without a doubt, a historic moment for our species, *Homo sapiens*, the moment we crossed a great distance, and we do not just mean the one that mediates between the Earth and the Moon. If you have seen the opening of Stanley Kubrick's *2001: A Space Odyssey*, where a primitive bone tool flies into space and is transformed into a space station, you will

---

[1] Regarding this chapter's title, "Pretext"—in both senses: "before the text" and motive.

The Women of the Moon. Daniel R. Altschuler Stern and Fernando J. Ballesteros Roselló.
© Daniel R. Altschuler Stern and Fernando J. Ballesteros Roselló 2019. Published in 2019 by Oxford University Press. DOI: 10.1093/oso/9780198844419.001.0001

know what we mean. (If you do not know, take this as a recommendation to see it.) It is estimated that some six hundred million people followed the adventure on television (when the world's population was about 3,626,000,000, half of today's). One of the authors bought a television set (black and white) for the landing, the other had to settle for listening through the belly of his mother. You can figure out which one was which. A total of nine missions manned by twenty-four astronauts left the grasp of Earth's orbit and went to the Moon, and twelve lucky humans lived the sublime adventure of walking on its surface, "a giant leap for mankind." Apollo 8 and Apollo 10 accomplished their missions of orbiting the moon; Apollo 13 could not land because of the accident immortalized in the words of its commander Jim Lovell (born 1928), the laconic: "Houston, we've had a problem here."

The last human to step on the Moon was Eugene Cernan (1934–2017), of Apollo 17, who followed his partner Harrison "Jack" Schmitt (born 1935) back into the lunar module on December 14, 1972. All the humans who have gone to the Moon have been male.

But our interest in the moon has a long history. Before Galileo pointed his small telescope at the Moon, marking the beginning of its scientific study, the Moon, the Sun, some wandering planet, and the occasional comet were the only celestial objects that seemed to *do* anything, and although nobody knew their real nature, at least they could be worshipped as gods. Of course, there were stars, fixed points of light that neither waxed and waned nor moved, and although people recognized the Milky Way, they had no clue what it really was. (It is our galaxy—from the Greek *gala*, meaning "milk"—seen edge-on.)

Skirting the line between science and symbolism, Dante Alighieri, in the *canto secondo del Paradiso*, of the *Divine Comedy*, written circa 1300, asks about the dark spots on the Moon—marring that symbol of purity:

> But tell me what those dark traces are
> Upon this body, which down there on earth
> Cause people to tell stories about Cain?
> (Cain, murderer of his brother Abel, was
> said to have been banished to the Moon,
> where he could be seen carrying a bundle
> of thorns to sacrifice.)

This was not entirely a rhetorical question: philosophers as well as poets tried to figure out why the stainless moon "smoothly polished, like a diamond" in Dante's words, had stains. The agreed solution was that, like a mirror, it reflected the imperfect Earth. Today we smile, but it was a clever way to understand the Moon in a manner that was consistent with other beliefs of the age.

Nevertheless, the Moon does reflect more than just sunlight. The nomenclature of lunar craters—1586 names honoring philosophers and scientists[2]—holds up a mirror to an important aspect of human history.

The geographic distribution of those honored reflects one facet of human history: the preponderance of Europe, and more recently the United States, in scientific and technical areas. Sorting lunar craters by the terrestrial geography of their namesakes, the United States, Germany, Great Britain, France, Russia, Italy, and ancient Greece (in order of importance), adds up to 1382 craters. There are only eight from South and Central America (five Argentinian, two Brazilian, and one Colombian), eight from Spain (after 1400), and only two from modern Africa (Willem Hendrik and Max Theiler).

The Moon reflects the intelligence and courage of some who have contributed to deciphering its history, studying it in detail, and even walking on its surface. But it also reflects the ignorance of others who believe utterly fanciful things about the Moon, including the idea that humans have never landed there!

Of those 1586 names—philosophers and scientists from Abbe to Zwicky—only 28 honor a woman. The atlas of the Moon reflects what was, and in many places still is, a negative view of women, a painful disregard for our own mothers.

The women of the Moon present us with an opportunity to meditate about all this, but perhaps more significantly, they offer us an opportunity to talk about their lives, mostly unknown today. Who were these women? Some were intellectual giants who triumphed in spite of everything; women we admire because they stubbornly refused to submit to the norms and prejudices of their time, acting in a way similar to Rosa Parks (1913–2005) who on December 1, 1955, refused to give up her seat to a white bus rider in Montgomery, Alabama. A few were sponsors of science and others, although not researchers themselves, were active in science communication. The most recent ones

---

[2] Elijah E. Cocks and Josiah C. Cocks (1995). *Who's Who on the Moon: A Biographical Dictionary of Lunar Nomenclature.* Tudor Publishers. This number does not include other craters named after gods or mythological beings, or with generic names officially unrelated to historical characters (such as Julienne, José, Kathleen... although unofficially some may have been inspired by a real person, as Kira, allegedly for the space scientist Kira Shingareva, but never recognized by the IAU).

represent a different kind of heroines, women who broke into the male world of space travel.

The women of the Moon tell us stories of love, sorrow, and courage, of some remarkable scientific achievements realized through perseverance, and of tragedies triggered by circumstance and prejudice. They offer us an opportunity to chronicle forgotten stories and review some well-known ones in a new light. They also give us an occasion to emphasize the injustice of 28 against 1558; there are abundant lunar craters still unnamed, and plenty of remarkable women who would deserve such an honor.

If, after reading this book, you go out one night with good binoculars to visit the women of the Moon (at least those visible), or if you are captivated by one of them and decide to read a full-length biography, we will be satisfied.

Understanding the Moon

Before we get to our women, we would like to provide you with a bit of background and some interesting, often misunderstood features of this unique body.

A Brief History of the Solar System

Let's begin with a summary to put the Moon in context[1]. About 4.6 billion (thousand million) years ago, a giant cloud—mostly hydrogen and helium, with a smattering of interstellar dust—contracted under its own gravitation. In its high-density center, the proto-Sun was formed, a huge spherical mass (in which hydrogen fusion had not yet begun). This mass, not quite yet a star, began to heat up. Around its center there developed a spinning disk of material that we call the solar nebula or protoplanetary disk. This disk extended to very distant regions, far beyond the present orbit of Neptune (which did not yet exist), and its density and temperature, very high near the center, decreased with distance. Different studies and possible scenarios provide a timescale of few million years for this period, short in comparison to Earth's age.

These circumstances determined to a large extent the chemical composition of the planets that would eventually be formed from the nebula's material by a process of aggregation. In denser areas, tiny ice-coated granules began to collide with each other until they adhered and formed ever-larger bodies. This process continued, eventually producing objects of several miles in size called planetesimals. These bodies had sufficient gravity to allow them to continue capturing material from the nebula and grow further, and faster. Like smaller particles, planetesimals occasionally collided, but the outcome of these collisions depended on the relative speed of the objects. If they came together

[1] For those readers who wish a more extended telling of this beautiful tale we recommend Dava Sobel's *The Planets*, Viking (2005).

The Women of the Moon. Daniel R. Altschuler Stern and Fernando J. Ballesteros Roselló.
© Daniel R. Altschuler Stern and Fernando J. Ballesteros Roselló 2019. Published in 2019 by Oxford University Press. DOI: 10.1093/oso/9780198844419.001.0001

slowly, they combined to form a body of greater volume and mass; if too fast, they fragmented into multiple pieces. Ultimately, there were several hundred bodies as large as the Moon in orbit around the newly formed Sun (now driven by fusion of protons at its hot center), and the collisions between them ended up giving rise to the planets we have today.

Far from the Sun, collisions occurred less frequently and less violently and the formation of planets took longer. The high temperature that prevailed near the young Sun evaporated the water and other volatile compounds that composed the cloud's granules so that in this region of the nebula the telluric planets were formed (from the Latin *Tellus*: Earth): Mercury, Venus, Earth, and Mars. Their appearance occurred in the relatively short time of several million years, during which the strong solar wind characteristic of young stars (a wind composed mainly of high-speed protons and electrons ejected from the Sun) was clearing the nebula around the Sun.

Further away, frozen water remained stable. There, two rocky spheres formed, and managed to attract material from the nebula before it dissipated, swept away by the solar wind. After several million years, the *gas giants* Jupiter and Saturn achieved their final form, with a nucleus similar to that of the terrestrial planets but surrounded by an enormous layer of hydrogen and helium, comprising about ninety percent of their mass, of composition not dissimilar to that of the original nebula or the Sun.

The formation of the *ice giants* Uranus and Neptune, with only twenty percent of their mass being hydrogen and helium, and the rest mostly water, methane, and ammonia, is not well understood, but it is thought that they formed closer to the Sun and later migrated to their current orbits. They are much less massive than the gas giants, but they still dwarf the inner planets. Those four giants comprise about ninety-nine percent of all mass orbiting the Sun.

Beyond Neptune, where the cloud of matter became vanishingly thin, collisions between planetesimals did not occur as often and so the aggregation process leading to planets never took hold. Pluto—discovered in 1930 by Clyde Tombaugh (1906–1997)—long considered the most distant planet, is not a true planet: and it is now accepted to be a survivor of the planetesimals that formed at the beginning. Pluto was always somewhat odd, a small object with only eighteen percent of the mass of our Moon, and an orbit about the Sun which is substantially different from those of the planets (with the highest eccentricity and inclination to the ecliptic—the plane along which the Earth orbits the Sun). Although Pluto has been stricken from the roll of the planets, it remains the brightest of thousands of objects that have recently been discovered beyond Neptune's orbit, in the so-called Kuiper Belt (a controversial designation[2]). Gerald Kuiper (1905–1973) was an astronomer of Dutch origin who in 1951, given the abundance of comets coming from regions beyond the orbit of Neptune, postulated the existence of a disk populated by millions of planetesimals in orbit around the Sun assuming that at those distances planetesimals would not have aggregated, an idea which had been proposed earlier by US astronomer Frederick Leonard (1896–1960) in 1930, shortly after the discovery of Pluto, and also by Irish engineer and astronomer Kenneth Edgeworth (1880–1972) in 1943. (Kuiper did think that the disk would have been dissipated by now.) The first Kuiper Belt object, 15760 Albion, was discovered by the Mauna Kea Observatory in Hawaii in 1992. The Belt extends up to fifty times the distance from Earth to the Sun (the astronomical unit, au).

Collisions between planetesimals gradually eliminated many of them from the inner regions of the nebula. Many more were thrown out of the nebula by the gravitational impulse they received when passing near one of the newly formed planets, and even if they were not "bounced" out of the nebula by the planets, these flybys may well have altered their orbits in ways that we do not know with certainty. Some of the planetesimals found a relatively stable place to remain until today. Between the orbits of Mars and Jupiter, for example, the asteroid belt formed, containing hundreds of thousands of small objects—the largest is known as Ceres, with a diameter of about 1000 km (620 miles)[3]. Ceres, like Pluto (2360 km or 1477 miles in diameter), belongs to a new class of objects, the "minor planets." Billions of planetesimals, flung from the central solar system, it is now thought, ended up forming a gigantic cloud in orbit around the Sun. Beginning at the furthest reaches of the Kuiper Belt and extending perhaps 100,000 au, this is known as the "Oort Cloud" in honor of Jan Oort (1900–1992), the Dutch

---

[2] International Comet Quarterly. What is improper about the term "Kuiper belt"? http://www.icq.eps.harvard.edu/kb.html

[3] Throughout the text we will use miles (mi) and kilometres interchangeably for the measure of things. Just note that 1 mile is 1.6 km.

astronomer who in 1950 deduced its existence from a study of the orbits of comets.

In these distant and frozen regions of the solar system, the material of the solar nebula was never altered by the Sun's heat or by collisions, and its very molecules escaped the violent events that gave rise to the planets. That is, the Oort Cloud and the Kuiper Belt form huge stores of planetesimals, an immense freezer where the original material of the formation of the solar system is conserved. If we were able to study these objects directly, we might learn a lot about the formation of the solar system, and from time to time we receive a free, although not unadulterated, sample.

These are meteorites[4], mostly small pieces (a few ounces) of different types of extraterrestrial material originating in the asteroid belt (with a very small fraction from Mars and the Moon), which, after surviving the high temperatures caused by crossing our atmosphere at high speeds (typically 45,000 mph), hit the surface of the Earth, leaving a crater if large enough.

NASA's New Horizons spacecraft, launched on January 16, 2006, flew by Pluto on July 2015 to provide us with the first extraordinary images of its surface and that of its moon, Charon. As we write it is speeding towards Ultima Thule, a small twenty-mile diameter body about four billion miles from Earth; for the first time we will see a Kuiper Belt object in any detail. New Horizons reached Ultima Thule on January 1, 2019, making it the farthest object ever to be visited by a spacecraft. By the time you read this, hopefully with fantastic images, our knowledge will certainly have been expanded, refined, and just possibly upended! Bringing the discussion back to Earth, before we return to our ultimate destination of the Moon, there are two dimensions which you'll need to wrap your head around in order to have a good handle on lunar geography: time and distance. The best estimates[5] of the age of the Earth—based on the time when Earth had achieved a stable size and shape—put it at around 4.6 billion (thousand million) years. Let's pause to consider this number, which is difficult to comprehend in a real, visceral, way. We can write the figure down, a four, a six, and eight zeros, and we can talk

---

[4] A "meteoroid" is any small body in outer space and if it enters our atmosphere it becomes a "meteor" or "shooting star." If it survives its passage through our atmosphere and reaches the ground it is called a "meteorite."

[5] These estimates are based on detailed studies of zirconium crystals ($ZrSiO_4$) and radiometric studies of meteorites and lunar rocks. We won't go into the details here.

about it in lectures. We can calculate with it and compare other times to it, but it is well-nigh impossible for us to *imagine* because it is difficult to relate it to our daily experience, which, unfortunately, is much shorter. With very good luck, our stay on this planet will last a hundred years; imagining even a million years is hard, and if a million years is hard, then nearly 5000 times that is inconceivable! If you could travel to the past in a time capsule, as happens in the movies, and go back one year every second, it would take six years to reach the Cretaceous and greet (from a safe distance) a *Tyrannosaurus rex*. It would take another 150 years to the time of Earth's formation. We are talking about a very, very long interval. Our species, *Homo sapiens*, has been around for some 200,000 years, a fairly long period in relation to our lives, but in geological terms we are newcomers. If we concentrated Earth's history in a three-hour-long movie, then our species would appear at the last second. If you blink, you might miss us.

Just as the human experience of time is infinitesimally small compared to the sweep of the formation of the planets, so our experience of space is almost useless in understanding the distances between celestial objects. Although in cosmic terms (that is to say, relative to the universe, or even the Milky Way galaxy) our solar system is a tiny place, a mere speck of microscopic dust, it has impossibly large dimensions on a human scale. Putting the scale of the solar system in terms of the fastest that most of us have traveled, imagine a trip by modern passenger jet. To make a complete voyage around the Earth we would have to travel by plane for about two days, and to reach the Moon we would need twenty days. Contemplate the Moon when it is visible during the day (it is a stunning image as it seems to be hanging there alone in the sky) and ponder while looking at it that you could reach it by plane in twenty days. If you continued traveling for another twenty *years*, you could reach the Sun. To visit Neptune, still within the confines of the planetary system, you would invest 600 years. If you wanted to continue to reach the nearest star, Proxima Centauri, it would take about 5 million years and don't think you would have traveled very far. Light, on the other hand, at its enormous speed, can cover that distance in only four years (and so we say that Proxima is four light-years away). The other stars you see on a dark night, as if they were diamonds on a dark cloth, might be hundreds of times further away. The universe is a Very Big Place—but let's return to the nearest astronomical body: our Moon.

The Moon and its Craters

Over the past century and a quarter, various hypotheses for the origin of the moon have—with theoretical and evidentiary support—been accepted in the scientific community. George Darwin (1845–1912; son of Charles) postulated, on the basis of study of both tidal dynamics and the three-body Earth–Moon–Sun gravitational interactions, that the moon was literally spun out of the earth—a young Earth, spinning much faster than it does today, flung off a chunk of crust (incidentally, creating the Pacific Ocean). This "fission" model competed through much of the twentieth century with the "capture" model, the theory that the moon was an already-existing body whose solar orbit brought it close enough to fall under the spell of Earth's gravity. Problems with both of these theories, particularly in light of the evidence of the moon rocks brought back to Earth on the Apollo missions, has brought us to a new consensus around the "great impact theory," which postulates that the Moon was formed about four and a half billion years ago as the result of a titanic collision between the barely formed Earth—indeed, Earth was still accreting matter from the solar nebula at the time of the collision—and a body the size of Mars. This object has been named Theia, after the Titan mother of Selene, the Moon's goddess, in Greek mythology.

The grand impact theory has recently been challenged and refined by new ideas that seek to better explain a few oddities (related to composition and isotopic similarities between the two bodies) of the Earth–Moon system. According to some planetologists, as a result of the beyond-massive impact between planet-sized objects, what would later become the Earth and the Moon started out as a huge (tens of thousands of kilometers), hot, and doughnut-shaped structure of rapidly spinning molten material named Synestia (meaning a structure held together), which later cooled to form the Moon and the Earth.

Perhaps the most important discovery of the lunar explorations was the verification that lunar rocks (a treasure of about 842 pounds) have a composition very similar to that of terrestrial rocks, but with less volatile materials (those with low melting points) and a greater proportion of refractory elements (those with high melting points), as if they were terrestrial rocks heated to very high temperatures, all of which supports the idea that the Moon was formed—in one way or another—as a result of the collision with Theia. Earth and its Moon, in short, are made of by and large the same material, but we take even our best theories about how this happened with a grain of salt and an open eye to new evidence.

The Moon's rotation on its axis is in synchrony with its orbit around the earth, which is in some senses an explanation—but in others simply a restatement—of the fact that the same side of the moon always faces Earth.

Since its formation, the Moon—with no atmosphere to protect its surface—has been struck by millions of space objects, leaving corresponding millions of craters, most small but a few huge ones. The Aitken impact basin (after Robert Grant Aitken, 1864–1951, American astronomer), on the non-visible side of the Moon, with a diameter of 2500 km and a depth of 12 km, is one of the largest impact structures in the solar system. On the visible side of the Moon, the largest crater is Bailly (Jean Sylvain Bailly, 1736–1793, astronomer and French revolutionary), located near the Moon's southwest end. Its diameter is 300 km and its depth about 4 km. The most conspicuous crater (see Figure 1) is the white-colored and star-shaped Tycho (Tycho Brahe, 1546–1601), 86 km in diameter, at the south of the Moon.

Today, after having examined the lunar rocks collected by the Apollo missions' astronauts, we know that the Moon's craters were created by impacts and not volcanic activity (although this was long considered a possibility). That they are the remains of astronomical impact is as expected and is consistent with the theory that planets arose from countless collisions among the various objects that formed in the solar nebula. The Moon has been geologically dead—there are no volcanoes, no earthquakes, no continental drift—for more than two billion years, and since it has no atmosphere or water to erode its surface, it offers an intact record of the history of this part of the solar system, holding many secrets about the time of planet formation and in particular the early eons of Earth's existence. On Earth, meanwhile, the action of water and air, as well as geological activity, has erased most of the wounds left by these impacts. That said, we *Homo sapiens* are the long-term consequence of one particular impact, that which caused the extinction of dinosaurs about 65 million years ago.

There are conspicuous differences between the two lunar hemispheres: the face we always see and the hidden face photographed by our lunar

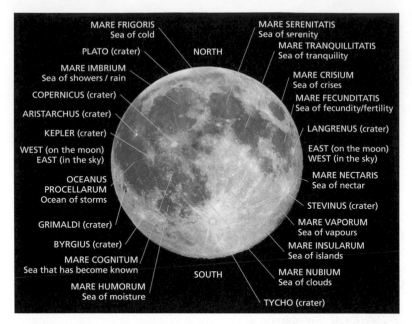

Figure 1 The lunar near side, showing lunar maria and some craters. By
Peter Freiman, Cmglee, Background photograph by Gregory H. Revera
[CC BY-SA 3.0
(https://creativecommons.org/licenses/by-sa/3.0)], from Wikimedia Commons. Modified by the
authors.

expeditions, crewed and robotic. The most striking is in the distribution
of "maria"—the plural of "mare" and the Latin word for "seas." These
dark, smooth-looking surfaces were at one time thought to contain water,
and we preserve the name for its mellifluousness, despite its inaccuracy.
Only one percent of the hidden face is covered by maria, while on the
Moon's visible side, maria cover about thirty percent. Maria are the
product of ancient magmatic emanations (not *exactly* volcanoes, but that
is a close earthly analog) that occurred over a billion years ago, covering the then-existing craters with lava. The best explanation for the difference between the two hemispheres connects it to radioactive elements
inside the Moon that produced the heat to generate lava. These elements
must have been concentrated closer to the visible surface, which was
therefore—to a great extent—filled with lava. Any theory of the Moon's
origins must reckon with this distribution of the maria, along with
everything else we know about lunar, terrestrial, and planetary rock
composition.

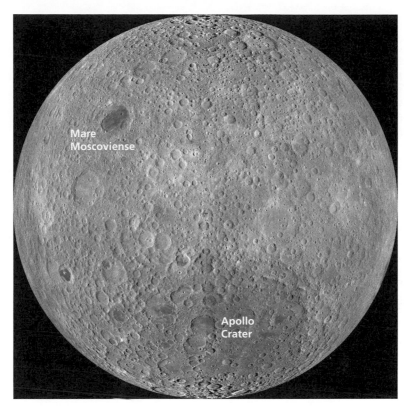

Figure 2 The lunar far side. The Aitken basin is the dark circular area that appears covering the lower quarter of the image. Mare Moscoviense is the prominent dark spot on the upper left, Apollo is almost opposite on the lower right. NASA/Goddard/Arizona State University.

Early Observations: van Eyck, Leonardo, Galileo, and Harriot

With the invention of the telescope, what Kepler called "lunar geography" was becoming possible. In the Somnium, he described the Moon as filled with mountains and valleys and as "porous as though dug through with hollow and continuous caves," a reference to the lunar craters Galileo had recently discovered with his first astronomical telescope.

CARL SAGAN[6]

[6] Carl Sagan and Ann Druyan (1980). *Cosmos*, p. 67. Random House.

If you look closely at the Moon, you see two things in it that are particular and that are not seen in the other stars. One is the shadow that it has, which is nothing but the rarity of its body, in which the rays of the Sun cannot be reflected as they do in the other parts; the other is the variation of its luminosity, because it is on one side, and then on the other, according to how the Sun sees it.

DANTE ALIGHIERI (c.1265–1321) (*Convivio*, Book II, Chapter XIII)

The first realistic representations of our satellite did not come from the hand of astronomers, but from the hands of artists[7]. Two small Moon drawings made by Leonardo da Vinci (1452–1519) around 1505 are one of the earliest pretelescopic extant portraits of our satellite. In them, we can appreciate the most important features, the dark areas or lunar maria and "the man in the Moon" (an example of "pareidolia," seeing recognizable images in essentially random visual information).

However, much earlier, in a painting of the crucifixion by Jan van Eyck (c.1385–1441) of 1425 we see a realistic representation of the Moon, much more detailed, even, than Leonardo's drawings. The Moon is shown in broad daylight, and although small, it is depicted in unprecedented detail. In any case, these early representations were made in pretelescopic times. When the first telescopes were invented and aimed at the Moon in the early seventeeth century, they truly opened up a new world.

(a)   (b)

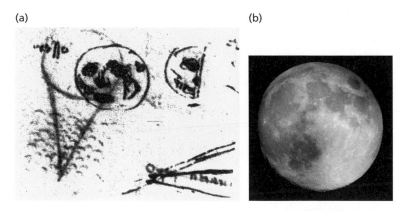

Figure 3 Comparison of Leonardo's sketches of the Moon (c.1505) and a photograph of the Moon by one of the authors.

[7] The authors wish to thank Bartolo Luque for allowing us to include material in the following section from "El año que la Tierra tuvo dos Lunas," published in *Astronomía*, II Época, No. 76, October 2005.

Figure 4 Jan van Eyck, *Crucifixion and Last Judgement diptych* (*c*.1430–1440). Oil on canvas, transferred from wood (the Moon is at the upper right). Currently at the New York Metropolitan Museum of Art, Fletcher Fund (acquired 1933).

In 1609, two men, from two different countries, had the privilege of seeing the Moon as no one had before, and through their eyes the Earth acquired two moons, each richer than the unmagnified sight that had been seen for millennia, but while one was larger, brighter, and *more* than had been seen before, the other truly represented not just a different image, but a different way of looking at the universe.

On November 30, 1609, Galileo Galilei deployed his twenty-magnification telescope in Padua, Italy. Many think of this event as the first telescopic observation of the heavens. But four months earlier, on July 26, Thomas Harriot (1560–1621) had preceded him in London with his six-magnification telescope. Harriot saw the old moon that poets and painters had been revering for millennia, the moon of Aristotle. Galileo saw a new moon, that of modern science. Fortunately, we have personal annotations that describe in detail what each man saw. How this happened is a story that illustrates how humans are able to interpret vague images to fit their expectations and beliefs. But, before presenting it, we need to briefly describe European "common knowledge" about the universe at the beginning of the seventeenth century.

In that Europe, still emerging from the Middle Ages, Aristotelian cosmology prevailed, which the Christian Churches had considered consonant with the Bible. In this view of the world, Earth and everything on it was composed of four elements—earth, air, water, and fire, as proposed by Greek philosopher Empedocles (490–430 BCE)—and, more importantly, all impurity or imperfection was limited to the Earth, the sublunary world, the center of the universe. In the sky were the celestial objects, not just revolving around the Earth, but composed of a fundamentally different material, the pure and perfect quintessence (literally, "fifth being"). The celestial bodies were separated from each other, meanwhile, by crystalline spheres, and the Moon, the closest object, was a perfect sphere antechamber of the superlunary world. The entire system

Figure 5   Albrecht Dürer, Virgin Mary with Child seated on a pillow on a crescent moon (c.1511), woodcut on paper, 20.6 × 19.6 cm.

remained surrounded by the immeasurable sphere of the stars, the last protective screen before reaching the Prime Mover. This distant region was easily assimilated ("Prime Mover" was an Aristotelian concept) by Christian mythology as Heaven, with a capital "H," where God and his heavenly court ruled. Western iconography had completely assimilated this image of the universe; for the Christian Church, the superlunary world was perfection.

Early Christianity, in its eagerness to recruit pagan devotees, associated the Virgin Mary with the popular hunting goddess Diana, also a virgin, and goddess of the Moon. In this way, by a quirk of fortune, the image of the Moon was united to that of Mary: the Moon, perfect, part of that Aristotelian superlunary world, became a symbol of the Immaculate Conception. Chapter XII of the Apocalypse begins with: "And there appeared a great wonder in heaven; a woman clothed with the sun, and the moon under her feet, and upon her head a crown of twelve stars" (Revelation 12 King James Version), and that image was reproduced in Christian iconography starting in the Middle Ages. A pure, crystalline, and immaculate moon, often in a rising phase or full moon, appeared as a symbol of the Virgin Mary's Immaculate Conception, as in the old engraving by Albrecht Dürer (1471–1528) shown in Figure 5[8].

But if the Moon is perfect, what of the spots and stains on it? This discussion, which we have briefly encountered already, has a long history. Already Plutarch (45–125 CE) in his dialogue "The Face of the Moon" presented a discussion about whether they were shadows, inhomogeneities of the crystalline density of the body, or terrain accidents as in the case of the earth's surface, while another character defended the notion that "the full moon is of all mirrors, in point of polish and of brilliancy the most beautiful and the most clear." Aristotle accepted the majority explanation: The Moon was an ideal sphere that reflected the perfection of the heavens. In fact, the Moon reflected everything. Scholars thought, therefore, that the Moon's ostensible marks were reflections of the Earth itself. What we were seeing in the heavens was

[8] Note in passing that the dogma according to which the Virgin Mary was preserved by God from original sin since her conception was proclaimed in 1854 by Pope Pius IX: "We declare, pronounce, and define that the doctrine which holds that the most Blessed Virgin Mary, in the first instance of her conception, by a singular grace and privilege granted by Almighty God, in view of the merits of Jesus Christ, the Savior of the human race, was preserved free from all stain of original sin, is a doctrine revealed by God and therefore to be believed firmly and constantly by all the faithful."

an image of our continents. An idea worthy of a genius, despite being erroneous and refuted by several, including Leonardo, who wrote the following[9]:

> Others say that the surface of the Moon is smooth and polished and that, like a mirror, it reflects in itself the image of our Earth. This view is also false, inasmuch as the land, where it is not covered with water, presents various aspects and forms. Hence when the Moon is in the east it would reflect different spots from those it would show when it is above us or in the west; now the spots on the Moon, as they are seen at Full Moon, never vary in the course of its motion over our hemisphere. A second reason is that an object reflected in a convex body takes up but a small portion of that body, as is proved in perspective.

Despite the logical naysayers like Leonardo, the idea of the Moon's purity and perfection was so rooted in the collective imagination that in 1609 instead of saying "pure as the driven snow" to refer to someone's innocence, one said: "pure as the Moon." With this context in mind, let us now turn to Harriot and Galileo.

Thomas Harriot[10]—mathematician, cartographer, and astronomer—was a singular character by the standards of our, or any, time. Although a pioneer of science in many fields, he did not publish most of his scientific works, for unknown reasons. His achievements remained hidden until his notebooks were rescued from the moths in the eighteenth century. It was then discovered that Harriot had attained important results in subjects as diverse as coding systems, algebra, spherical geometry, and kinematics. The great mathematician Friedrich Bessel used Harriot's 1607 observations of Comet Halley almost two centuries later to calculate the orbit of the famous object. He was a master of optics, about which he maintained an abundant correspondence with Johannes Kepler. He came up with the correct explanation of the formation of the rainbow and discovered, before Descartes and Snell, the law describing the refraction of light. Today, however, we know this mathematical relationship as Snell's law and not as Harriot's law. It has been said of him that he was the most pioneering scientist you've never heard of; one of the best examples of the oft-repeated admonition: "publish or perish."

[9] Gibson Reaves and Carlo Pedretti (1987). Leonardo da Vinci's drawings of the surface features of the moon. *Journal for the History of Astronomy*, Vol. 18, No. 1/Feb, p. 55.

[10] Allan Chapman (2009). A new perceived reality: Thomas Harriot's Moon maps. *Astronomy & Geophysics*, Vol. 50, Issue 1, February 1, pp. 1.27–1.33. https://doi.org/10.1111/j.1468-4004.2009.50127.x

Harriot probably invented the telescope independently. Not only is it true that his lunar observations preceded Galileo's, but from his manuscripts historians have been able to determine that he was also the first observer of Jupiter's moons (which he called new planets), and of sunspots. On July 26, 1609, he pointed his six-magnification telescope to the Moon. We have the crude drawing he made of that first famous observation. In it, there are clearly some spots and the *terminator*, the line that separates the area of light from the area of darkness on the moon's surface.

The drawing is incorrect in the sense that the terminator must intersect the satellite's shape at two diametrically opposed points. But what interests us most is the detail that the terminator appears as a broken and irregular line. Harriot, an Aristotelian and convinced of the purity of the Moon like most scholars of the time, must have been surprised by such a vision but could offer no explanation. Harriot "knew" that the Moon was immaculate and crystal clear, so he was unable to interpret his observation. In the words of science historian Gerald Holton[11]:

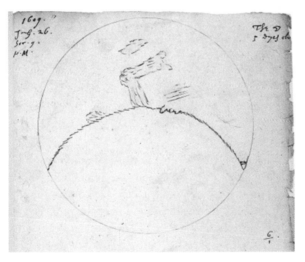

Figure 6 Harriot's July 26, 1609 sketch of the Moon—the first known drawing of a telescopic body—approximately six inches in diameter with the Mare Crisium shaded at the top. Reproduced by kind permission of Lord Egremont, Petworth House Archives HMC 241/9 fol 26. West Sussex Record Office, Chichester.

[11] Gerald Holton (1996). On the art of scientific imagination. *Daedalus*, Vol. 125, No. 2, p. 186.

"Harriot sees, but the current presuppositions make it difficult for him to undertake the intellectual transformation, to cross from sense experience to a new way of understanding." And so it came to be in 1609, in England, that although something strange happened to the terminator, the Moon was still smooth and perfect.

Galileo Galilei (1564–1642), meanwhile, was an enthusiastic follower of Copernicus's heliocentric model, which proposed that the Sun was the center of the universe. He had devoted his efforts so far to lay the foundations of modern kinematics (the study of the motion of objects without regard to their causes) in open opposition also to Aristotelian physics. But in 1609, with the telescope to his eye, he finally found "evidence for the senses" of the Copernican truth. On November 30, four months after Harriot, he turned his twenty-magnification telescope to observe the Moon. The light that came through his telescope would illuminate modern science. His famous watercolors of the moon's phases in those observations also show the terminator as a broken and sinuous line. But in addition to this evidence from the border between light and dark, the lunar surface itself is clearly rough and earthy. In this way, while from England the traditional Moon was still seen, in Italy, Galileo, observed another very different one: a new Moon, in many aspects surprisingly similar to Earth.

Galileo published his revolutionary astronomical discoveries in 1610 in the book *Sidereus Nuncius*, the messenger of the stars. There he described the Moon's surface, Jupiter's satellites, the phases of Venus, and sunspots. It was a universe radically different from the Aristotelian one that had prevailed for more than 2000 years. What Galileo saw and understood about the Moon is described with detail and precision: "it is not smooth, uniform and exactly spherical . . . but irregular, rough, and full of cavities and prominences, similar to the face of the Earth, dressed in mountain chains and deep valleys." Galileo not only sees but understands that there are no qualitative differences between the Moon and the Earth. In fact, although in the drawings of *Sidereus Nuncius* the maria depiction is quite accurate, Galileo exaggerates, probably on purpose, the lunar topography. Some historians think that the famous crater Bohemia (the huge crater that divides the terminator in Galileo's imagery) is an exaggeration with didactic spirit, since it does not, in fact, exist. Let us read the words as written by the great man himself[12]:

[12] Galileo (1610). *Sidereus Nuncius*. Venice. https://www.gutenberg.org/files/46036/46036-h/46036-h.htm

Figure 7 Galileo's drawings of the Moon in Siderius Nuncius.

Let me speak first of the surface of the Moon, which is turned towards us. For the sake of being understood more easily, I distinguish two parts in it, which I call respectively the brighter and the darker. The brighter part seems to surround and pervade the whole hemisphere; but the darker part, like a sort of cloud, discolors the Moon's surface and makes it appear covered with spots. Now these spots, as they are somewhat dark and of considerable size, are plain to everyone, and every age has seen them, wherefore I shall call them great or ancient spots, to distinguish them from other spots, smaller in size, but so thickly scattered that they sprinkle the whole surface of the Moon, but especially the brighter portion of it. These spots have never been observed by anyone before me; and from my observations of them, often repeated, I have been led to that opinion which I have expressed, namely, that I feel sure that the surface of the Moon is not perfectly smooth, free from inequalities and exactly spherical, as a large school of philosophers considers with regard to the Moon and the other heavenly bodies, but that, on the contrary, it is full of inequalities, uneven, full of hollows and protuberances, just like the surface of the Earth itself, which is varied everywhere by lofty mountains and deep valleys.

The appearances from which we may gather these conclusions are of the following nature:—On the fourth or fifth day after new moon, when the Moon presents itself to us with bright horns, the boundary which divides the part in shadow from the enlightened part does not extend continuously in an ellipse, as would happen in the case of a perfectly spherical body, but it is marked out by an irregular, uneven, and very wavy line, as represented in the figure given, for several bright excrescences, as they may be called, extend beyond the boundary of light and shadow into the dark part, and on the other hand, pieces of shadow encroach upon the light:—nay, even a great quantity of small blackish spots, altogether separated from the dark part, sprinkle everywhere almost the whole space which is at the time flooded with the Sun's light, with the exception of that part alone which is occupied by the great and ancient spots. I have noticed that the small spots just mentioned have this common characteristic always and, in every case, that they have the dark part towards the Sun's position, and on the side away from the Sun they have brighter boundaries, as if they were crowned with shining summits. Now we have an appearance quite similar on the Earth about sunrise, when we behold the valleys, not yet flooded with light, but the mountains surrounding them on the side opposite to the Sun already ablaze with the splendor of his beams; and just as the shadows in the hollows of the Earth diminish in size as the Sun rises higher, so also these spots on the Moon lose their blackness as the illuminated part grows larger and larger. Again, not only are the boundaries of light and shadow in the Moon seen to be uneven and sinuous, but—and this produces still greater astonishment—there appear very many bright points within the darkened portion of the Moon, altogether divided and broken off from the illuminated tract, and separated from it by no inconsiderable interval, which, after a little while, gradually increase in size and brightness, and after an hour or two become joined on to the rest of the bright portion, now become somewhat larger; but in the meantime others, one here and another there, shooting up as if growing, are lighted up within the shaded portion, increase in size, and at last are linked on to the same luminous surface, now still more extended. An example of this is given in the same figure. Now, is it not the case on the Earth before sunrise, that while the level plain is still in shadow, the peaks of the loftiest mountains are illuminated by the Sun's rays? After a little while does not the light spread further, while the middle and larger parts of those mountains are becoming illuminated; and at length, when the Sun has risen, do not the illuminated parts of the plains and hills join together? The grandeur, however, of such prominences and depressions in the Moon seems to surpass both in magnitude and extent the ruggedness of the Earth's surface, as I shall hereafter show. And here I cannot refrain from mentioning what a remarkable spectacle I observed while the Moon was rapidly approaching her first quarter, a representation of which is given in the same illustration, placed opposite page 16. A protuberance of the shadow, of great size, indented the illuminated part in the neighborhood of the lower cusp; and when I had observed this indentation longer, and had seen that it was dark throughout, at length, after about two hours, a bright peak began to arise a little below the middle of the depression; this by degrees increased, and presented a triangular shape, but was as yet quite detached and separated from the illuminated surface. Soon around it three other small points began to shine, until, when the Moon was just about to set, that triangular figure, having now extended and widened, began to be connected with the rest of the illuminated part, and, still girt with the three bright peaks already mentioned, suddenly burst into the indentation of shadow like a vast promontory of light.

What made both of them look at the Moon with such different eyes (or should we say minds)? Without doubt, the worldview that they brought to their viewing of the Moon played a part: Aristotelian for Harriot, and Copernican for Galileo. Two geniuses, two leading characters of scientific rigor, saw the same thing and interpreted what they saw in totally different ways due to their prejudices. But there is more. When we look at the drawings of what they had seen, we cannot help but notice the artistic abyss that separates them. Did the different artistic gifts of these scientists matter?

Harriot's England of the early seventeenth century was that of William Shakespeare (1564–1616) and Ben Jonson (1572–1637), where words triumphed over paintings. In artistic terms, England was worlds away from Renaissance Italy. Galileo had been hired at the age of twenty-five as a professor of mathematics at the Academy of Design in Florence, the most important center in the world for painting and architecture. There he taught geometry and in 1613 he was elected as a member of the Academy. His mathematical work complemented not only his scientific work but also the artistic labors of his colleagues; he studied and taught how shadows project on different surfaces, mastering the art of perspective. Coming from this milieu, Galileo was able to interpret in a bold way that the terminator's irregularities were actually shadows produced by the roughness of the lunar terrain, and to definitively conclude that there were valleys and mountains on the Moon. From the shadows (a problem of inverted optics where the object that produces the shadow is reconstructed), he was even able to estimate the height of some lunar mountains at 6000 meters: higher than the Alps, something hard to believe for the intellectuals of the time. Not only was Galileo's *scientific* imagination ready to receive the images of a new Moon, but also his *visual* imagination would provide a more fertile field than Harriot's for new ideas to take root.

The later history of *Sidereus Nuncius* is well known; while Galileo did not immediately pursue the ideas, his *Dialogue Concerning the Two Chief World Systems* (published in 1632) set Aristotelian geocentrism and heliocentrism (with imperfection in the celestial bodies) in direct contrast. Although this latter book initially enjoyed the imprimatur of ecclesiastical censorship, Galileo was accused of heresy, and, after a humiliating trial in which false evidence was presented, he was sentenced to house arrest, where he spent the rest of his days. The book was immediately included in the Church's "anti-hit parade"—The Index of Prohibited Books of the Holy Inquisition (*Index Librorum Prohibitorum*)—from which it was not removed until 1835. And it was only in 1992, that the Church, lamenting "the errors committed by both parties," in the words of Pope John Paul II, "admitted" its error with Galileo. Despite this, the triumph of *Sidereus Nuncius* was unstoppable. It was the triumph of reason, of science, and of the counterculture, since it was soon distributed in all scientific circles in Europe. It was followed by *Considerations and Mathematical Demonstrations on Two New Sciences*, a book that Galileo got published in Leiden, smuggled from the prison of his house in Florence.

The influence of *Sidereus Nuncius* also reached art and in a short time, Galileo's observations had made their way back to England in an art form they could reckon with: poets like Milton, Dryden, and Donne were writing about a new moon with steep mountains. Religious art also appropriated the new Moon, and gradually the old pure and crystalline sphere disappeared. The birth of the new Moon would be initially reflected in the arts by the Florentine painter Ludovico Cardi, called Cigoli (1559–1613), a friend and admirer of Galileo. In his last work, *Immacolata*—a fresco at Santa Maria Maggiore in Rome, representing the ascension of the Virgin Mary standing on a moon—it was no longer Harriot's Moon,

Figure 8 Fresco by Cigoli in the Basilica di Santa María Maggiore, Rome (painted in 1612). It is the first representation of the "blemished" Moon with craters (the moon appears deformed because of perspective, the photograph having been taken from the ground). On the right, a painting by Murillo of 1660 also showing an "imperfect" moon.

but Galileo's. In the iconographic representations of the Virgin after Galileo, moons at the feet of Mary begin to appear with imperfections, craters, and seas (*Maria*, by the way).

Galileo was the first man conscious of witnessing a lunar sunrise, the Sun rising above the Moon's mountains, as he describes his observations. Harriot, to be sure, *saw* it first, but he was unable to understand it. In July 1610, after reading *Sidereus Nuncius*, Harriot again pointed his telescope to the Moon. He had seen through his veil. In the drawing of his new observation, lunar craters are clearly visible. The Earth once again had only one Moon. There is a third character to this story: Simon

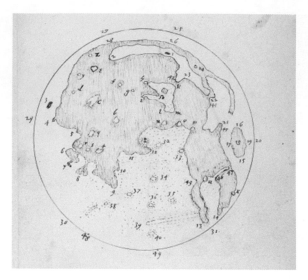

Figure 9 Harriot's Moon, 1610. Reproduced by kind permission of Lord Egremont, Petworth House Archives HMC 241/9 fol 30. West Sussex Record Office, Chichester.

Marius (1573–1625), a German astronomer. Although we do not know the exact dates of his observations, he also gazed through a telescope, and in 1614 published *Mundus Iovialis*, containing his observations of Jupiter and its moons, starting—he claimed—in 1608. This raised some controversy about who did what first, and in 1623 Galileo accused him of plagiarism. But historical evidence shows that he did indeed work independently. He writes[13]: "I have come to be certain that these stars have Jupiter for their acknowledged center, and are borne around him, exactly as Mercury, Venus, Mars, Jupiter, and Saturn move around the Sun as their center." Marius was also the first to observe and describe the Andromeda nebula (galaxy). He was honored by the International Astronomical Union (IAU) with a crater on the Moon.

This case was resolved by history: the moons are known collectively as the Galilean moons but their individual names (Io, Europa, Ganymede, and Callisto) are those proposed by Marius, after the mistresses of Jupiter (a literary touch suggested to him by Kepler).

[13] A.O. Prickard (1916). The "Mundus Jovialis" of Simon Marius. *The Observatory*, Vol. 39, pp. 403–12. http://adsabs.harvard.edu/abs/1916Obs....39..403

Figure 10 John Adams Whipple, The Moon, August 6, 1851, daguerreotype. Collection of Middlebury College Museum of Art, Vermont. Purchased with funds provided by the Christian A. Johnson Memorial Fund and the Overbrook Foundation, 1989.009.

In closing, we show one of several early photographs (a daguerreotype) of the Moon (Figure 10), this one taken on August 6, 1851 by John Adams Whipple (1822–1891), obtained using the eight-inch Harvard Observatory refractor assisted by George P. Bond (1825–1865), son of the observatory director. (We know that John William Draper (1811–1882) (father of Henry Draper whom we will meet later) obtained images of the moon in 1840 but they have unfortunately not been preserved.) One of Whipple's and Bond's images of the Moon was awarded the prize for technical excellence in photography at the Great Exhibition held at the Crystal Palace at Hyde Park in London in 1851.

The Moon's Size and Distance

Here we first wish to point out that the determination of our Moon's distance and size is a problem first tackled by Greek scholars with results which, considering their rudimentary instrumentation, were quite good. We will then briefly tell you how we go about this in modern times.

Aristarchus of Samos (310–230 BCE) thought that the Sun was at the center of the universe and that the Earth orbited the Sun and turned on its axis almost 2000 years before Nicolaus Copernicus (1473–1543). Unfortunately, his manuscripts (except for his only extant work: "On the sizes and distances to the sun and moon"[14]) have been lost and we only know about his heliocentric ideas from a mention in a book by his younger contemporary, Archimedes (c.287—c.212 BCE), who writes about Aristarchus: "His hypotheses are that the fixed stars and the sun remain unmoved, that the earth revolves about the sun in the circumference of a circle, the sun lying in the middle of the orbit."

Using the right triangle formed by the Earth, the Moon, and the Sun when, as seen from the Earth, one half of the Moon is illuminated by the Sun, he tried to measure their angles and thus deduce their distances trigonometrically. The instrumental limitations of the time did not allow him to get an accurate result (he estimated eighty-seven degrees whereas the real angle is eighty-nine degrees and fifty minutes) and he found that the Moon was about twenty terrestrial radiuses (actually it is sixty) away and that the Sun was twenty times further away than the Moon (and twenty times larger since their angular sizes are almost equal), when in fact that factor is 390. Aristarchus also used the geometry of a lunar eclipse (when the Earth interposes itself between the Sun and the Moon) to estimate the relative sizes of the Earth and the Moon, by measuring the radius of the shadow of the earth on the Moon. He thus found that the Earth was about three times the size of the Moon and the Sun much greater than either. From this fact alone, it did not make sense that it was the Sun that orbited the Earth—it must be the other way around.

Regarding the size of the Moon, if the size of the Earth were known it would follow from the eclipse measurements. Eratosthenes of Cyrene (c.276—c.194 BCE) (a good friend of Archimedes), working in Alexandria, determined the terrestrial circumference (assuming a spherical earth and the Sun at a great distance), in a famous experiment some sixty years later. He obtained the size of Earth's circumference as 250,000 stades (a measure of length estimated to be between 578 and 685 feet), and therefore its circumference as between 27,400 and 32,500 miles, which is quite a good estimate (the real value of meridional circumference is 24,860 miles)[15]. More than the accuracy of his measurements, what is of importance is the method he used. Once the size of the Moon is known,

---

[14] Thomas Heath (1913). *Aristarchus of Samos* (Oxford University Press academic monograph reprints).

[15] Newlyn Walkup (2005). Eratosthenes and the mystery of the stades. Mathematical Association of America. *Convergence* Vol. 2. https://www.maa.org/book/export/html/116342

its distance can be obtained from its angular size in the sky. Figure 11 illustrates the relationship between the Earth and the Moon to scale.

Hipparchus of Nicaea (*c*.190—*c*.120 BCE) improved the results of Aristarchus. Now recognized as the first star-cataloger (he assembled a catalog of 1024 stars published in the Almagest of Ptolemy) and the inventor of trigonometry, Hipparchus discovered the phenomenon of the precession of the equinoxes and, using the geometry of lunar eclipses, calculated that the distance to the Moon was sixty-six times the terrestrial radius, a result very close to the true one (about sixty). Not bad for doing this 2000 years ago.

Figure 12, an image created by astronomer Anthony Ayiomamitis from Athens, illustrates his method. It is a superposition of several photographs taken during the lunar eclipse of August 16–17, 2008. The Earth's shadow on the Moon is clearly visible, and from it the relative size of the Moon and the Earth can be estimated. If you use the image and your tools from geometry class (compasses and straightedge), you can approximate that the Earth is three times larger than the Moon. There are a few sources of error in this method, to be fair: in addition to the fuzzy edge of the earth's shadow, it is assumed that the shadow is the same size as the Earth, which is not true since the Sun's rays are not strictly parallel; because the Sun is larger than the Earth, its shadow tapers off with distance and is only three-quarters as wide at the Moon's distance. If we use the modern value of the Earth's diameter of 7918 mi, then we can conclude that the diameter of the Moon is 2639 mi (the real value is 2158 mi). Knowing that the Moon subtends about half a degree (0.5 degrees) in the sky, we get its distance (assuming a circular lunar orbit—another approximation) from the proportion: $2639/0.5 = 2\pi D/360$, resulting in $D = 302{,}407$ mi. (Average distance is 239,000 mi). Not bad for such a rough calculation.

In more recent times, increasingly precise measurements of the Moon's distance have been obtained by measuring lunar parallax (another form of triangulation). The idea of measuring the

Figure 11 The Earth–Moon system to scale.

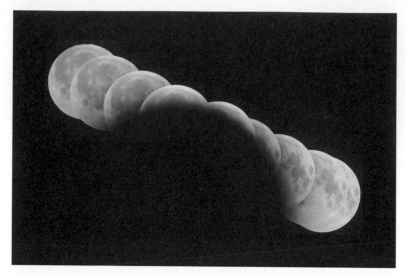

Figure 12 The shadow of Earth during a lunar eclipse. Photo by Anthony Ayiomamitis.

distance to the Moon by parallax is simple: from two points on Earth as widely separated as possible (say, an observatory in Berlin or Greenwich and another one in Cape Town), the angular position of a lunar feature (e.g., the center of a crater) is measured simultaneously.

Then, knowing the distance that separates the two observation points, a triangle is constructed with the lunar geographic feature as its third point. With one known distance and two observed angles, it is possible to calculate the length of the missing sides. In practice, this is very difficult, since an error of a few minutes of arc (sixtieths of a degree) in measurement produces an error of tens of thousands of kilometers, to which we should add the error due to the uncertainty in the distance between the two observation points (see Chapter 3 on 'Nicole-Reine de la Briere Lepaute').

Modern techniques for measuring the Moon's distance are based not on trigonometry, but instead use laser retroreflectors placed on the Moon's surface by the Apollo astronauts and by Soviet unmanned missions. By shining a laser beam from the Earth on these mirrors and measuring the light's round-trip time, measurements of the distance can be obtained with a precision of a few millimeters. The most surprising result may not be the distance itself, but the fact that the Moon slowly moves away from Earth by 3.8 centimeters per year. It does not seem like much but over the scale of cosmological time, it adds up: a mere 500 million years ago (about one-ninth the age of the Earth), the Moon was 19,000 km closer.

The "Dark Side" of the Moon

Careful observers have known for centuries that the Moon always shows us the same face, that is to say, that the Moon's period of rotation and the orbital period around the Earth are equal. This is not an accident; it's due, rather, to the same tidal forces that cause the seas on earth to rise and fall (roughly) twice a day. Tidal forces (see 'Appendix B') distort any body in a gravitational field, stretching it along the line that joins their centers. When the Moon was younger, it did in fact rotate faster than it orbited around the Earth (it was also closer, and so the tidal amplitude was even greater than it is today), and so it experienced tides. Because there were no bodies of water to absorb this energy, the tides occurred in the rock itself, deforming the Moon as you might squeeze a ball of clay between your hands. This deformation (which, as the body rotates, is continually squishing the body in a new direction) heated the Moon, just as you can observe if you take a wire and bend it back and forth repeatedly (this is a nice illustration of conservation of energy: the work of bending the wire is turned into heat energy). Much of this heat energy dissipates into space. But it can build up! The tides generated by Jupiter on its nearest moon Io are enormous, resulting in deformations of its surface hundreds of meters high, and causing its interior to heat to such an extent that Io has volcanos.

The reciprocal effect of tidal generation on Earth and the Moon has had consequences for the rotation and orbital characteristics of both bodies. Given that the total energy of the Earth–Moon system that is not lost as heat is conserved, the end result is that the orbital and rotational energy must decrease to compensate for that which has dissipated as heat over billions of years. In the Moon's case, this caused its rotation around the Earth and around its axis to slow until it became synchronized with the Earth's gravitational pull, so that today the terrain raised by the tidal force is always the same and there are no more friction losses.

On Earth, the tides are experienced as oceanic bulges, not aligned with the Earth–Moon direction because the Earth is spinning more or less twenty-eight times faster than the Moon's orbit. This causes friction between the ocean floor and the slow-moving ocean bulges, which acts to slow down the Earth's rotation. Our most precise atomic clocks do in fact show that the length of the day increases fifteen microseconds each year. It does not seem like much, but in a million years the day will last about twenty-eight hours and, looking in the other direction, the Cambrian day, 500 million years ago, lasted twenty-one hours. At the same time, as we have mentioned, the Moon moves away from earth by a few centimeters a year, to conserve angular momentum.

The above is highly simplified, and the detailed study of Earth–Moon dynamics is complex and mathematically difficult, but it does not change the essence of the process. One of the pioneers with these calculations

Figure 13 "Earthrise" stamp and photo. The postage stamp was issued on May 5, 1969, and also displays a minor cosmological error. USPS and NASA.

was the mathematician George Howard Darwin (1845–1912), son of Charles Darwin. We met him briefly as we discussed lunar origins.

While the same side of the Moon is always visible from Earth, the reverse is not true. The photo of the Earth's light and dark sides taken[16] by the Apollo 8 astronaut William Anders (born 1933), circulated all over the world and implanted in the minds of many an inescapable consciousness of the Earth's finitude and solitude, in Jim Lovell's (who orbited the Moon on Apollo 8 and Apollo 13) words: "a grand oasis in the vastness of space." The photo is known as "Earthrise" by analogy to the Moon's rise in the terrestrial sky, but this is wrong. As the Moon always shows us the same face, the Earth in the lunar sky is fixed.

Furthermore, it is common to refer to the Moon's hemisphere that we do not see as "the dark side," another errant concept that entered the popular imagination to the point where one of Pink Floyd's best-selling albums was called "The Dark Side of the Moon."

But if you think about it a little, there is no such thing. With respect to the Sun, the Moon rotates on its axis and the side lit by sunlight changes with time. When we see the Moon in its phases, the half of the moon in sunshine is partially on the hemisphere that faces Earth, and partly on the "far side." When we have a new moon the non-visible (or "far-side") hemisphere is full of light. One of the women of the Moon whom you'll soon meet, Mary Somerville, describes it in the following way in her *Mechanism of the Heavens*:

> [T]he earth, which must be so splendid an object to one lunar hemisphere, will be forever veiled from the other. On account of these circumstances, the remoter hemisphere of the moon has its day a fortnight long, and a night of the same duration not even enlightened by a moon, while the favored side is illuminated by the reflection of the earth during its long night. A moon exhibiting a surface thirteen times larger than ours, with all the varieties of clouds, land, and water coming successively into view, would be a splendid object to a lunar traveler in a journey to his antipodes.

To end with another thing that might puzzle you, there are only very few instances in which you get a second full moon in one calendar month (about once every two and a half years). This leads to the

---

[16] On 70 mm Kodak Ekta chrome film, using a modified Hasselblad camera, in orbit around the Moon on December 24, 1968.

expression "Once in a blue Moon" to indicate a rare event. It is not clear why "blue" is used, although it is a rare event when the Moon is actually seen bluish due to unusual atmospheric conditions (caused by volcanic eruptions or large forest fires). There will be a blue moon on October 31, 2020 (first full moon on October 1), and the next one will be on August 31, 2023.

Lunar Nomenclature

We can trace the history of lunar nomenclature to Plutarch of Chaeronea (c.45–120 CE). He was, as far as we know, the first to put names to the places on the Moon. Observing what seemed to be a large featureless area on the visible side, he thought that it must be a plain destined for the souls of the dead, the plain of Persephone (daughter of Zeus and the harvest goddess Demeter, and queen of the underworld); he also believed there was a similar plain on the far side, the Elysian plain (a kind of paradise). The idea that the dark spots visible on the Moon must be seas came from Plutarch. When comparing the Moon with the Earth he wrote[1]: "but like as our earth has deep and great gulfs— one of them flowing inwards towards us through the Pillars of Hercules; others flowing outwards as the Caspian, and those in the Red Sea—in like manner there are deep places and gulf-like in the moon."

During the Middle Ages, these phrases generated the belief that the dark spots were indeed seas like ours, and at least one of the names of the terrestrial seas mentioned by Plutarch, the Caspian Sea, inexplicably became part of the lunar geography, at least for a while. Indeed, three seventeenth-century selenographers, the Dutchman Michael van Langren (1598–1675), better known as Langrenus, Thomas Harriot (whom we have already met, although his map was not well known until it was published in 1965), and the Frenchman Pierre Gassendi (1592–1655), independently named the same structure as Mare Caspian; which seems to indicate that it was a name habitually used in seventeenth-century Europe to designate this lunar region. (We now know that area as Mare Crisium—the "Sea of Crises.")

It was just in that century when the era of selenography began, so-called by analogy with geography (Selene, as already mentioned, was the Greek Moon goddess, and Gea or Gaia, depending on Greek dialect,

---

[1] Plutarch. "On the apparent face in the orb of the moon." https://people.sc.fsu.edu/~dduke/lectures/plutarch-moonface.pdf

The Women of the Moon. Daniel R. Altschuler Stern and Fernando J. Ballesteros Roselló.
© Daniel R. Altschuler Stern and Fernando J. Ballesteros Roselló 2019. Published in 2019 by Oxford University Press. DOI: 10.1093/oso/9780198844419.001.0001

was the goddess of the Earth). Almost coincidentally, several astronomers embarked on this task[2]. The times were propitious; exploration and mapping of the New World, was en vogue, and there was ample work for cartographers. The recent invention of the telescope facilitated the discovery on the Moon of a very complex world that also had to be mapped. A practical earthly problem, accurately determining geographic longitude (one of the most complex problems in the history of astronomy), also put pressure on scientists to turn their eyes to the Moon[3]. In the golden age of sailing, the solution to this problem required knowing the time in a known (and distant) place, so that by comparing the time of local sunrise, sunset, or noon with the time of these events at one's place of departure, a simple calculation (of fifteen degrees of earthly rotation per hour) would reveal one's position. It was thought that the Moon could play the role of a universal clock, visible from all over the world: measuring the position of the lunar terminator (the line of separation between the part of the Moon illuminated by the Sun and the part in shadow), and observing which craters or mountains were crossed at each moment by said terminator, together with some precise tables, could show the time with an accuracy better than a minute. For this, it was necessary to make an accurate map of the Moon and name its structures.

Gassendi was possibly the first to prepare, in telescopic times, a systematic nomenclature of the lunar landscape, although rudimentary and referring only to the larger structures. It did not get published, and has only survived to the present in his personal writings.

The nomenclature of Langrenus, geographer of King Felipe IV of Spain, was widespread and widely used by other astronomers. As a flattering courtier, he named the lunar features with the names of European royalty and nobility—Philippi IV, Mare Austriacum, or Mare Borbonicum—but also with the names of philosophers, scientists, and explorers. Of his nomenclature, only the craters Endymion (a young shepherd loved by Selene) and Pythagoras have survived to the present, together with the crater Langrenus, which he dedicated to himself. This is because shortly after him, two other works were published that

---

[2] For a thorough and entertaining tale, of which this chapter is just a summary, we recommend: Ewen A. Whitaker (2003). *Mapping and naming the moon: a history of lunar cartography and nomenclature.* Cambridge University Press.

[3] Dava Sobel. (1995). *Longitude.* Fourth Estate.

completely eclipsed any previous lunar cartography: those by Hevelius and Riccioli.

The Polish astronomer Johannes Hevelius (1611–1687) published in 1647 his magnum opus, *Selenographia* (only two years after the work of Langrenus), in which he named all the seas, mountains, and lunar structures (there was still no mention of craters) that he could see with his telescope. He initially thought of naming these structures in honor of illustrious people, but to avoid possible misgivings, he changed his mind and decided to take inspiration from terrestrial geography itself. The names are sonorous and long, Chersonesus Taurica, Lacus Hyperboreus Inferior, Celenorum Tumulus . . ., and were quickly adopted by selenographers from all over Europe.

His competitor was the Italian Jesuit astronomer Giovanni Battista Riccioli (1598–1671), who published another influential work only four years after the *Selenographia* of Hevelius: the *Almagestum Novum*. Riccioli had no problem in filling the lunar topography with names of illustrious persons, in most cases related to astronomy. The lunar structures that we now call craters were named after ancient and modern philosophers and astronomers such as Archimedes, Aristarchus, Aristotle, Copernicus, Plato, Tycho, Eratosthenes, Kepler, Galileo, and Alphonsus (for Spanish King Alfonso X the wise), or Grimaldi, his disciple. The Mutus crater, located near the lunar south pole, was named in recognition of the Mallorcan astronomer and military engineer Vicente Mut Armengol (1614–1687), and Munosius was dedicated to the sixteenth-century Valencian astronomer Jerónimo Muñoz (c.1520–1591). (This we mention with some regional pride: one of the authors is Valencian). The seas and lands received names with a medieval astrological flavor, such as the Sea of Crisis, the Sea of Rains, the Sea of Nectar, the Sea of Fertility, and the Sea of Tranquility.

Riccioli's distribution of names is not random. In his scheme, he divided the moon into eight sectors and populated them with related characters. We have, for example, an Alexandrian sector where we can find different figures of antiquity who lived in that city.

Riccioli was a supporter of the cosmological system of Danish astronomer Tycho Brahe (1546–1601), a curious mixture between a geocentric and a heliocentric system in which all the planets, except the Earth (which was not considered to be a planet) and the Moon, revolved around the Sun; the Sun and the Moon, by contrast, revolved around the Earth. Tycho Brahe's system may seem capricious, but Brahe was no

fool, nor was he simply kowtowing to religious authority. His measurements told him without any doubt that the planets revolved around
the Sun; but, on the other hand, he knew that if the Earth also did this,
then the stars seen from opposite sides of our orbit should show a
change in their apparent position (called "parallax") and Brahe was
not able to measure this. He concluded, therefore, that the Earth did
not move[4].

For Riccioli, Brahe's cosmological system was perfect, because it not
only kept the Earth at the center of the universe, as the Church advo-
cated, but explained all known astronomical observations of planetary
movements. In his lunar atlas, therefore, he put all the astronomers
who argued that the Sun was the center of the universe in the Ocean
of Storms (Oceanus Procellarum—the largest of the maria) "so they
would drown." There we find the craters Copernicus, Kepler, Galileo,
and Aristarchus, among other heliocentric scientists. On the other
hand, he named the most resplendent and visible crater on the Moon,
a crater that has bright traces of *ejecta*, products of the impact that
formed it[5], crossing wide areas of the Moon like white rays, "Tycho."
The idea is clear: Tycho's light of truth illuminates the other astronomers
(who are wrong).

For almost two hundred years, the cartographies of Hevelius and
Riccioli remained in force. In a long-term terminological standoff, it
became standard practice for any published lunar map to include both
nomenclatures, and two names were given for each lunar structure.
Luckily for Riccioli (and sadly for Hevelius), his were shorter. When a
lunar map of small size had to be printed—for example, in a book—
usually the Riccioli nomenclature was used. Over time, this small
advantage became important, and by the mid nineteenth century,
Riccioli's nomenclature had become the standard. It is, in fact, the basis
of the current nomenclature. Of Hevelius's, only ten names of mountains are conserved today, four in the same place where he placed them,

---

[4] Unfortunately for Brahe, the correct explanation is not that the Earth does not
move, but that the stars are much further away than he supposed, and their parallax is
therefore much smaller. It was not until the year 1838 that Friedrich Wilhelm Bessel
(1784–1846) measured for the first time a stellar parallax (0.314 seconds of arc for the star
61 Cygni) and demonstrated convincingly that the Earth does revolve around the Sun.
On the other hand, given that his system kept him in good standing with the Church,
this may have been a fortunate error.

[5] These bright rays of ejecta show that this is a very recent crater.

and six in places changed by later selenographers[6]. A large crater of 118 km on the western edge of Oceanus Procellarum bears his name.

Over the nineteenth century, telescopes became better and cheaper, and selenography underwent a new advance: an army of contributors to knowledge. Amateurs by now had access to tools that even the professionals of previous centuries could not have imagined, and with these, they were discovering new lunar features, unknown to Riccioli and Hevelius. This new generation of selenographers hastened to baptize these features, expanding Riccioli's nomenclature. Each selenographer used their own names and it all led to a plethora of new nomenclatures (almost as many as available telescopes). Among their works we must mention those of Wilhelm Lohrmann (1796–1840) in 1824, William R. Birt (1804–1881) and John Lee (1783–1866) for the British Association in 1865, and Johann N. Krieger (1865–1902) in 1898, since some of our women of the Moon owe their craters to them[7].

Not all extensions of Riccioli's nomenclature had equal prestige. At the beginning of the twentieth century, three were most popular among astronomers: that of Johann H. Mädler (1794–1874) and Wilhelm Beer (1797–1850), of 1837, a complete and monumental work that also introduced the current custom of cataloging with letters the smaller craters around named craters (e.g. the Tycho B crater, west of Tycho's crater); the 1876 volume by Edmund Neison (1849–1940), based on Mädler's work, extended and published in an extremely popular book, and that of Julius Schmidt (1825–1884), of 1878, another very complete work of lunar mapping that was born, in fact, to compete with Mädler's map. With rare exceptions, any textbook from the early twentieth century that presented a detailed lunar map used one of these three nomenclatures.

For many astronomers, this situation was unsustainable; they wanted a standardized and internationally agreed nomenclature for lunar structures. This movement was led by the British mathematician and selenographer Samuel A. Saunder (1852–1921), who described the confused state of lunar nomenclature to the Royal Astronomical Society. After a debate within the society, during which even a radical

---

[6] Those centuries of use did not leave the maps of Riccioli and Hevelius untouched. Over time, they were filled with errors of copy, misprints, positional changes, additions and inexplicable subtractions, and other changes that have persisted until today.

[7] For his part, Johann H. Schröter stands out for being the first to introduce the word crater in 1787 to name the most common structure on the Moon.

change of nomenclature was proposed—starting from scratch—it was finally agreed to keep what was already well established and to start from a comparative study of the nomenclatures of Mädler, Neison, and Schmidt, finding and codifying the points of agreement and resolving the contradictions. The person in charge of this work, as we will see, was Mary Adela Blagg, one of our women of the Moon. After several years comparing the work of these selenographers, and under the direction of Saunder, she finally published in 1913 her *Collated List of Lunar Formations*.

This document was immediately influential but it soon became definitive. The International Astronomical Union (IAU) was founded in 1919 at a meeting of astronomers in Rome, in large part to serve as the international authority regarding the nomenclature of celestial bodies and their surface structures. Lest this aim seem trivial, nomenclature serves as the baseline for international scientific communication. One of its first decisions was to approve Blagg's work as a base document and appoint her to the lunar nomenclature commission of the IAU. In later years, Blagg and the lunar commission polished their work to produce a new document, accepted by the entire international community: *Named Lunar Formations* (1935). Any crater named since has been an addition to this base document, and any proposed changes have been arbitrated by the IAU. Its policy for the Moon is stated in the most recent version of the list: the craters are designated with the names of "Deceased scientists and polar explorers who have made outstanding or fundamental contributions to their field. Deceased Russian cosmonauts are commemorated by craters in and around Mare Moscoviense. Deceased American astronauts are commemorated by craters in and around the crater Apollo. Appropriate locations will be provided in the future for other space-faring nations should they also suffer fatalities. First names are used for small craters of special interest."

That definitive statement was not reached without controversy, however. When lunar exploration began in earnest with the US and Soviet space programs, the discovery of new structures that needed to be named grew exponentially: the new ships saw the Moon closer, and small structures became more and more important. In 1964, the Ranger 7 probe was launched by NASA to crash on a small, unnamed mare on the visible side. The lunar commission of the IAU decided that this mare should receive a proper name and since it was going to be known very well, it was named Mare Cognitum, the "known sea."

After the first probes circumnavigated the Moon (the Soviet Luna 3 was the first, in 1959) and started to photograph and map its unnamed and unknown far side, which was very different from the visible face, the opportunities for an expanded nomenclature multiplied. There were few marias and many craters. The Soviets considered themselves in a certain way "owners" of the far side of the Moon that their ships had discovered and to the 1967 meeting of the IAU, held in Prague, they brought an immense photographic mosaic, based on photos from the Luna 3 and Zond probes, so that people could "walk on the Moon." This map was accompanied by a list of names of which almost half were Russian—and among those, the Russian cosmonauts were given prominence. This policy of fait accompli did not please the lunar commission in particular, nor the IAU in general.

After years of tense negotiations, a consensus was reached that was embodied at the next meeting of the IAU (held in Córdoba, Argentina, in 1970): the system of nomenclature would remain consistent with the one already used on the visible side, but, recognizing that new probes would soon be exploring other worlds in the solar system, it could be extended in passing to name the features of other planets and celestial bodies. However, several points of the Soviet proposal were accepted, such as extending the nomenclature not only to scientists but also to astronauts and cosmonauts. The previous resolution of 1961 did not allow naming structures after living people, but in the case of the pioneers of space exploration, an exception was made. Thus, nine eponyms from each side of the Iron Curtain would be accepted, six of them dead and three still alive. The Russian cosmonauts would be grouped in the vicinity of Mare Moscoviense, on the far side, while of the American astronauts, some are near Tranquility Base (the site of the Apollo 11 landing) on the visible side, and the rest are near the Apollo impact basin, located, like Mare Moscoviense, on the far side of the Moon. It was specified that, if there were other nations with space travelers, appropriate locations would be identified.

As we have mentioned, of the 1586 craters on the Moon named after individuals (others have generic or descriptive names), only 28 of those names belong to women. It is estimated that there are 300,000 craters larger than 1 km in diameter on the Moon's visible side alone—that is, there are enough to remedy the injustice. A single example should suffice to illustrate the situation: there is Sklodowska, named in honor of Marie Skłodowska Curie (Nobel Prizes in Physics, 1903, and Chemistry,

1911), and Curie, a crater in honor of Pierre Curie, her husband and collaborator. There is also a Joliot crater, officially named by the IAU in 1961 in honor of Frédéric Joliot-Curie, husband of Irène Joliot-Curie, Marie and Pierre's daughter. In 1935, Irène and Frédéric jointly received the Nobel Prize in Chemistry in recognition of their synthesis of new radioactive elements. But there is no crater for Irène, for unknown reasons.

A Joliot-Curie crater does exist on Venus, named in honor of Irène by the IAU in 1991. But in our opinion, a crater on the Moon is not the same as one on Venus or Mars: for those of the Moon, at least on the visible side, anyone with a small telescope can see the memorial.

Although Sappho appears as the name of a lunar crater in many indices, the IAU indicates the following: "This name was never approved by the IAU. The approved name for this feature is Stark V (25.1 S/133.3 E/25 km)." Johannes Stark (1874–1957) won the Nobel Prize in Physics in 1919 but later became a supporter of the Nazis, so much so that he was tried and sentenced to four years in prison after the war. We would have preferred Sappho, author of the following lines:

> The sinking moon has left the sky,
> The Pleiades have also gone.
> Midnight comes – and goes, the hours fly
> And solitary still, I lie.

Sappho ($c.630/612$–580 BCE) was a Greek lyric poet who was born and lived on the island of Lesbos, near the Turkish coast, in the Aegean Sea. She enjoyed great fame in antiquity—Plato referred to her two centuries after her death as the "tenth muse." Little is known about her life; what little we do know is gleaned from the fragments of her poetry that have survived, or filtered through ancient historical tradition to such an extent that it is impossible to distinguish reality from myth. This is not particular to Sappho; reality and legend are mixed together in most biographical accounts of ancient authors. Even her poetry survives almost entirely in quotations—possibly misquotations—in the work of others, or translations. Nor is it known when she died, but in her late poems, she describes herself as an old woman who enjoys a quiet, poor life, in harmony with nature.

The only thing that seems to be certain is that she lived in Lesbos, and that her poetry had a strong erotic component, in particular, a suggestion of love for other women: "Sleeping on the chest of a tender

companion," for example. (This is the origin of the term "lesbianism.")
Another fragment also shows her attention to the Moon:

> The stars about the full moon
> lose their bright beauty when she, almost full,
> illumines all earth with silver.

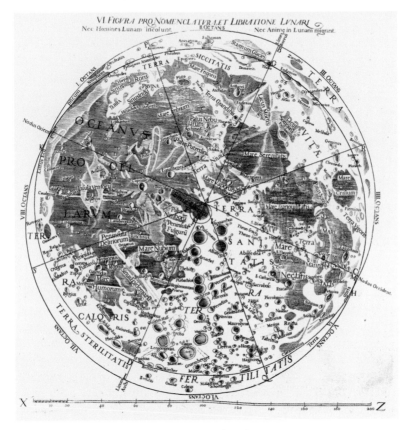

Figure 14 Lunar Map by Riccioli (1651), Almagestum Novum.

WOMEN OF THE MOON

Now rose the moon, full and argentine,
While round stood the maidens, as at a shrine.

SAPPHO

The women of the Moon form an eclectic group; no uniform criteria
are discernible for their recognition. A few are well-known figures in
the history of science, such as Marie Curie and perhaps Gerty Cori; the
modern astronauts, too, might be familiar, but most are quite unknown
today. They include Nobel Prize winners[1], less well-known scientists—
mostly astronomers—and more recently astronauts who lost their
lives exploring space. But those twenty-eight names are there for a reason, and each has a story to tell. If we exclude the two most ancient
ones—Hypatia and Catharina—(if indeed they were two, as we shall
explore), and the five most recent ones (astronauts), then the rest were
born between 1723 and 1896. Only one is alive at the time of this writing:
Valentina Vladímirovna Tereshkova (born 1937).

---

[1] Only 49 of the 893 Nobel Prizes awarded until 2018 (not counting 27 organizations)
have been awarded to women.

The Women of the Moon. Daniel R. Altschuler Stern and Fernando J. Ballesteros Roselló.
© Daniel R. Altschuler Stern and Fernando J. Ballesteros Roselló 2019. Published in 2019
by Oxford University Press. DOI: 10.1093/oso/9780198844419.001.0001

The twenty-eight women of the moon (in order of appearance)

Name[2]	Data	Year[3]	Diam. (km)	Lat.	Long.	Visible?[4]
Hypatia, of Alexandria	Ancient Egypt, mathematician (355 or 370–415)	1935	41×28	−4.25	22.58	Yes
Catharina, Catherine of Alexandria	Ancient Greece, theologian, philosopher (c.287–c.370)	1935	100	−17.98	23.55	Yes
Lepaute, Nicole-Reine de la Brière	France, astronomer (1723–1788)	1935	17	−33.3	−33.69	Yes
Herschel, Caroline Lucretia	Great Britain, astronomer (1750–1848)	1935	13	34.48	−31.29	Yes
Somerville, Mary Fairfax Greig	Scotland, physicist, mathematician (1780–1872)	1976	15	−8.33	64.96	Yes
Sheepshanks, Anne	Great Britain, philanthropist (1789–1876)	1935	25	59.24	17.04	Yes
Bruce, Catherine Wolfe	US, philanthropist (1816–1900)	1935	7	1.16	0.37	Yes
Mitchell, Maria	US, astronomer (1818–1889)	1935	30	49.77	20.17	Yes
Clerke, Agnes Mary	Great Britain, astronomer (1842–1907)	1973	7	21.68	29.8	Yes

[2] Lunar craters are referred to simply by one name: Hypatia, Catharina, Lepaute, etc. In this column, the name by which the crater is known appears first, before the comma.

[3] This is the year in which the IAU approved the appointment.

[4] Whether the crater is on the visible side of the moon. Whether the crater is visible with the naked eye depends on atmospheric visibility and visual acuity but in general, craters bigger than 40 miles (only a few) can be discerned with the naked eye on a clear night, and craters larger than 1 mile can be seen through a small (eight-inch) home telescope.

Name	Data	Year	Diam. (km)	Lat.	Long.	Visible?
Kovalévskaya, Sofia Vasílyevna	Russia, mathematician (1850–1891)	1970	115	30.86	−129.44	No
Maunder, Annie Scott Dill Russell[5]	Great Britain, astronomer (1851–1928)	1970	55	−14.52	−93.88	No[6]
Fleming, Williamina Paton[7]	US, astronomer (1857–1911)	1970	106	14.91	109.28	No
Cannon, Annie Jump	US, astronomer (1863–1941)	1964	58	19.88	81.36	Yes
Maury, Antonia Caetana de Paiva Pereira[8]	US, astronomer (1866–1952)	1935	18	37.11	39.69	Yes
Leavitt, Henrietta Swan	US, astronomer (1868–1921)	1970	66	−44.86	−139.89	No
Blagg, Mary Adela	Great Britain, astronomer (1858–1944)	1935	5	1.22	1.46	Yes
Proctor, Mary	US, astronomer (1862–1957)	1935	53	−46.43	−5.04	Yes
Skłodowska-Curie, Marie[9]	Poland, physicist, Nobel Prize (1867–1934)	1961	127	−18.04	96.15	No

(Continued)

[5] The Maunder crater is named jointly for Annie and Edward W. Maunder (1851–1928), her husband and scientific collaborator.

[6] Maunder and Sklodowska, although on the back side, are so close to the edge that they can be visible from Earth thanks to lunar libration (the phenomenon by which, due to eccentricities in the moon's orbit and the daily rotation of the earth, an observer at one spot on Earth can see more than half of the Moon's surface—about fifty-nine per cent).

[7] Commemorating Williamina Paton Fleming together with Alexander Fleming (1881–1955), doctor, discoverer of penicillin, and Nobel laureate—and no relation to Williamina.

[8] Named for Antonia Maury and Matthew Fontaine Maury (1806–1873), Antonia's cousin, astronomer and oceanographer.

[9] The crater named simply "Curie" is in honor of Pierre.

Name	Data	Year	Diam. (km)	Lat.	Long.	Visible?
Meitner, Lise	Austria, physicist (1878–1968)	1970	87	−10.87	113.1	No
Noether, Amalie Emmy	Germany, mathematician (1882–1935)	1970	67	66.35	−114.18	No
Jenkins, Louise Freeland	US, astronomer (1888–1970)	1982	38	0.37	78.04	Yes
Bok, Priscilla Fairfield	US, astronomer (1896–1975)	1979	45	−20.26	−171.58	No
Radnitz Cori, Gerty Theresa	Czechoslovakia– US, physiologist, Nobel Prize (1896–1957)	1970	65	−50.48	−152.91	No
Astronauts/ cosmonauts						
Resnik, Judith Arlene	US, Challenger crew (1949–1986)	1988	20	−34.15	−150.84	No
McAuliffe, Sharon Christa	US, Challenger crew (1948–1986)	1988	19	−33.24	−149.77	No
Chawla, Kalpana	US, Columbia crew (1962–2003)	2006	14	−42.8	−147.5	No
Clark, Laurel Blair Salton	US, Columbia crew (1961–2003)	2006	15	−43.7	−147.7	No
Tereshkova, Valentina Vladímirovna Nikolayeva	Soviet Union, Vostok 6 cosmonaut (1937–)	1970	31	28.21	143.84	No

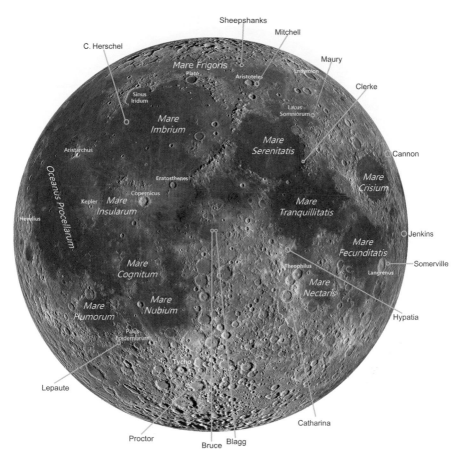

Figure 15 Finding chart for craters on the visible side of the Moon (LRO image and the authors).

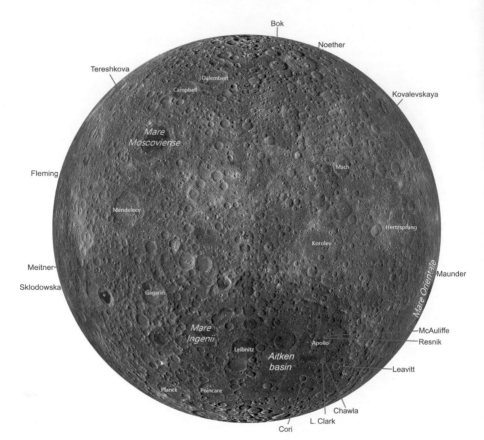

Figure 16 Finding chart for craters on the far side of the Moon, for completeness sake since you will not be able to see most of them, although those near the rim (Sklodowska, Maunder) are sometimes visible due to the "libration" or wobble of the Moon due to its orbital characteristics.

1

Hypatia of Alexandria
(355 or 370–415)

Figure 17 Gilded mummy portrait of a woman, often referred to as Hypatia. From Er-Rubayat, Egypt. Roman Period, *c.*160–170 CE.

There was a woman at Alexandria named Hypatia, daughter of the philosopher Theon, who made such attainments in literature and science, as to far surpass all the philosophers of her own time. Having succeeded to the school of Plato and Plotinus, she explained the principles of philosophy to her auditors, many of whom came from a distance to receive her instructions.

SOCRATES OF CONSTANTINOPLE, Christian Greek historian and contemporary of Hypatia. From his *Historia Ecclesiastica*[1], Book VI, chapter 15

Hypatia of Alexandria[2] was a mathematician, astronomer, and philosopher; she was born in Alexandria (perhaps obviously), daughter of the mathematician and astronomer Theon, the last director of the Library at Alexandria. We do not know the exact date of her birth; two years are attested in the historical tradition: 355 and 370 CE. This should not be surprising since the date of birth of a person in antiquity had in general little relevance, except to

---

[1] http://go.uv.es/ferbaro/scolasticus

[2] We recommend the film "Agora" (2009), directed by Alejandro Amenábar, starring Rachel Weisz as Hypatia, although controversy exists about its historical accuracy. (After all it is a *movie*.)

The Women of the Moon. Daniel R. Altschuler Stern and Fernando J. Ballesteros Roselló.
© Daniel R. Altschuler Stern and Fernando J. Ballesteros Roselló 2019. Published in 2019 by Oxford University Press. DOI: 10.1093/oso/9780198844419.001.0001

mark when she came of age. The year of her death, however, is well known because its tragic circumstances left a lasting memory: 415 CE. Her sad end is often associated (not entirely precisely) with the destruction of the Library at Alexandria, the jewel of the Ancient World, although in fact the Library, after suffering a series of catastrophes throughout its long history, had been dismantled twenty-four years before her death. Her story cannot be understood without the story of the Library, and so we must turn back the clock some 650 years.

The great Library at Alexandria was founded, together with its sister institution, the Museum (from the Greek *mouseion*, meaning "seat of the muses"), around 300 BCE by Ptolemy I, successor to the Egyptian part of Alexander the Great's empire and therefore the first pharaoh of the Hellenistic period in Egypt. The foundation of the Library was part of an active policy (which Ptolemy and his successors would consistently follow) of support for the arts and sciences. The Museum, despite its name, had little to do with what we now associate with the word. Inspired by the Lyceum of Aristotle, it was a place dedicated to the collection of knowledge, research, and teaching. In short, it had characteristics that did not return until the first universities were founded in the Middle Ages, and for all intents and purposes, we can think of the Museum as a university. The Library was, in fact, not a stand-alone institution but rather a part of the Museum, storing (and producing on demand) the books necessary for the academic functioning of the numerous scholars who worked in the Museum, paid directly by the State. The Library of Alexandria was the largest in antiquity and had an immense collection of books; at its peak, it came to hold almost a million volumes.

One of its first directors, Eratosthenes (c.276 BCE—c.195 BCE), was an astronomer and geographer, best known as the first person to calculate the circumference of the Earth. He provided the institution with an astronomical observatory, which he installed in one of the Library's terraces (in 230 BCE). This observatory of Alexandria would be the most important in the Ancient World, and it was there that the two most renowned astronomers of antiquity made their discoveries: Hipparchus of Nicaea (c.190—c.120 BCE) discovered the precession of the equinoxes and made the first catalog of stars of which we have evidence; and Claudius Ptolemy (c.100—c.170 CE) assembled his influential treatise, the *Almagest*. This work would provide the dominant framework for understanding the relationship of earth and sky for a millennium and a half: the so-called Ptolemaic system, with the Earth at the center of the universe and the planets orbiting it. Lest we underestimate the sophistication of Claudius Ptolemy's thought, he understood that simple circular motion would not describe his observations, and so the Ptolemaic system proposed that planets orbited around points placed on these circles around the Earth. This cosmological system would dominate the astronomy of the Middle Ages until Copernicus, Kepler, and Galileo overthrew it.

Museum, Library, and observatory were in the royal district of Brucheum, where the monarchy and royal officials resided. But there was another, auxiliary, library in the peripheral neighborhood of Racotis, inside the temple of Serapis (Serapeum). All of these institutions existed within a cultural and political climate that was dominated by a singular fact: the ruling caste of Egypt, since Alexander conquered it and his general (then Pharaoh) Ptolemy I Soter (c.367—283 BCE) inherited it, were Macedonian Greeks, regarded by the Egyptian natives as foreign invaders and received with understandable misgivings. Ptolemy I, however, endeavored to be accepted by his people. To unify the Greeks and Egyptians in his realm, he put religion in the service of politics and invented a god based on the popular Egyptian gods Osiris and Apis, but of clear Greek aspect: Serapis. Ptolemy devoted himself to extending the cult of Serapis as the patron deity of his reign and as a unifying element for Egypt. Ptolemy I's temple in honor of Serapis was, however, modest and it fell to his grandson, Ptolemy III, to expand the Serapeum to its final magnificent form.

This expanded Serapeum was soon used not only as a cult temple but also as an institution for dissemination of Hellenic culture and indeed conversion to Hellenic ideals among the population: it received books from the Library of Alexandria to translate from Greek into Egyptian, to be used in the teaching of the Egyptian popular classes. Much as the Museum functioned as a university, the Serapeum served as a school. Over the centuries, the Serapeum acquired more books and served as a secondary warehouse for the mother library, which was running short on space: new acquisitions as well as working copies were filling its shelves, and copies of many books were transferred to the Serapeum.

Three hundred years after its founding, the great Library suffered its first serious mishap. The reign of Cleopatra, last pharaoh of the Hellenistic dynasty (and great-great-great-great-great-great granddaughter of Ptolemy I) was a tumultuous one. Both Rome and Egypt were undergoing civil strife, and Julius Caesar, in 48 BCE, entered Egyptian territory chasing Pompey. Egypt, already in a civil war over the succession to its throne fell in line with the Roman sides, with Cleopatra lending her support to Caesar in exchange for his help in securing the Egyptian throne. Her brother (and competitor for the throne) Ptolemy XIII, meanwhile, aligned himself with Pompey. Caesar was besieged in Alexandria by supporters of Ptolemy XIII and to defend himself, he ordered the ships of the port to be set on fire. Although the classical sources are confusing and contradictory, it seems certain that this fire spread to some port warehouses that belonged to the Museum and where several books were stored which were destined for the Library, as well as some quantity of blank papyrus. The Library itself was not damaged, but the rumor mill inflated the burning of the books to the point that many classical authors (even to the authority of Seneca and Plutarch) associated this incident in the port of Alexandria with the Library's disappearance. Egypt would ultimately emerge from the chaos as a Roman province.

Despite this setback, the Library remained in operation for three additional centuries (now under Roman imperial funding), although suffering the vicissitudes of the various wars that, from time to time, ravaged the region. Among them, we mention the invasion of Alexandria in 272 CE by the kingdom of Palmyra, which had declared war on the Roman Empire. During the recovery of the city by the Romans, the royal district of Brucheum, where the museum and Library were located, was practically destroyed. Probably both buildings were greatly damaged. If there was any attempt to rebuild them, it was short-lived, because only twenty-five years later Alexandria itself revolted against the Empire, an uprising crushed by Emperor Diocletian. The Museum's buildings and the Library were destroyed. But the Serapeum, which was in a peripheral neighborhood, survived. Everything that could be saved from the Library and the Museum was taken to the Serapeum, and there the Museum continued its work.

The Serapeum thus became the only center of scientific research of the entire Roman Empire, but its operation was encountering more and more difficulties. By the third and fourth centuries CE, Christianity was spreading throughout the Empire, and institutions devoted to other deities were at risk when political leaders adopted the new religion. Furthermore, the old distrust on the part of the Egyptian populace of their Greco-Roman rulers had never fully disappeared, and during the fourth century, the Church became a channel for these nationalist sentiments with large swathes of the population adopting the nascent Christian religion as a way of differentiating themselves from their rulers. In this way, and especially in politically peripheral areas such as Alexandria, the classical world and everything associated with it were stigmatized with the label of "pagan," and that included science: the Egyptian people stopped feeling that the Serapeum belonged to them.

During the reign of Emperor Constantine (whose mother was Christian), Christian religion was legalized (in 313 CE) and the first council of the Church, at Nicaea (in 325 CE) took place. That same year, a Roman decree was approved by which the salaries of the Serapeum's faculty were discontinued, forcing them to give private classes in their homes or set up private academies to survive. This was the state of the Empire when Hypatia's father, Theon, assumed the Serapeum's leadership. The classical world declined and science in Alexandria was performing its swan song.

Theon was a renowned astronomer and mathematician. He predicted and observed various eclipses, and wrote several books, including a treatise on astrolabes. But the work he is best known for today was the production of editions of science books. They were called "commentaries" and included the author's original text, as well as running comments from the editor (making clear the authorship of each part of the text), which in many cases exceeded the original text. They also sometimes included exercises, diagrams, and demonstrations[3].

Theon published several commentaries of classic works. Of special importance was his commentary on Euclid's *Elements* (another former user of the Museum, six centuries before). As any student of mathematics knows, Euclid is considered the father of geometry, and the thirteen volumes of his *Elements* were the basis of mathematics for centuries. It became, according to some counts, the second most printed book after the Bible since the printing press was invented, and remained compulsory at many universities until well into the nineteenth century. So widely distributed was Theon's commentary on

---

[3] A current example can be seen in the book *Intelligent Life in the Universe*, initially published in Russian by Iosif Shklovsky, in 1962, and then published in English along with Carl Sagan in 1966. Sagan extends the original text, duplicates it, and marks with different typefaces the original text of Shklovsky and his own comments.

Euclid that while no copies of Euclid's work without commentary survive, it is now known through Theon's edition[4].

It was into this distinguished household—and this world in conflict—that Hypatia was born, the first woman in history known to have dedicated herself professionally to science. With the name he gave to her (Hypatia roughly translates as "the greatest"), it should not be surprising that Theon put all his interest in her receiving a thorough and rigorous education, both intellectual and physical. While the majority of women of her time were dedicated to the care of the home and children, and had no access to education, Hypatia trained with the main brains of the museum and learned mathematics and astronomy. She completed her training in Italy and Athens, studying philosophy. In addition, following the Greek tradition of *sophrosyne* (excellence of mind and body brought about by temperance, prudence, purity, and self-control), she submitted her body to great physical discipline: she exercised daily and led a life of chastity, keeping herself a virgin as a form of control over her own body (in fact, she never married)[5]. Theon's education paid off and Hypatia became an intelligent and beautiful woman, according to all testimonies. Skilled in mathematics and astronomy, in rhetoric and philosophy, she soon outdid her own father in knowledge.

Although some authors have referred to her as the last director of the Library of Alexandria, it seems that Hypatia was never directly related to the Serapeum, except for the fact that the director was her father and that she knew all the academics of this institution. An adherent of the philosophical system of the Neoplatonic school[6], she set up her own academy in her home, where she taught philosophy classes. Her students were drawn from the highest classes of Alexandria, both

---

[4] This was the case until the nineteenth century, when a copy was found in the Vatican archives that did not derive from Theon's commentary, but from a text of Byzantine origin.

[5] Although historian Damascio (Damascius, *The Philosophical History*, ed. & trans. P. Athanassiadi [Athens 1999]) cites the philosopher Isidor of Alexandria as a possible husband of Hypatia, they were not in fact contemporaries. When Isidor was born, Hypatia was already dead.

[6] Hypatia's Neoplatonism derives from Plato's theory of forms that aspires to a spiritual world and sees material reality as a shadow of that world. Neoplatonism uses abstraction from individual instances to reach the Platonic forms (such as truth, beauty, etc.), a process which eventually leads the adept to the One, the underlying principle of all nature. Mathematics plays a special role for Neoplatonists of the fourth century as the nature of mathematics is to abstract—to derive *ideas* from material things.

Christian and pagan, and it is clear from surviving documents that the group maintained important bonds of friendship with each other throughout her life. Many of them would later reach important positions; among them the Christian Synesius of Cyrene (c.373–c.414), who later became bishop of Ptolemais (Libya) and with whom she maintained a deep friendship through her life (many of the letters they exchanged are preserved; they are one of the main sources of information about the life of Hypatia); and Orestes, who would become the imperial governor of the province of Egypt. Her students, both during the time they spent in her school and later throughout her life, addressed her as "Lady," "the Lady," "my Lady," . . ., which may indicate that she was older than them, and might hint that the correct date of her birth was 355, the earlier of the two attested dates. (Another common nickname by which her students referred to her was "The Philosopher.") The good relationship she had with this important group endowed her with a notable influence in political life; Synesius himself, by then a bishop, often asked for advice in his letters.[7]

Alas, very little remains of Hypatia's scientific work, but it must have been significant because it was widely commented on by other authors. Because of this, we know that she wrote a treatise on astronomy, the *Astronomical Canon*, and she carried out a review of Claudius Ptolemy's astronomical tables. Like her father, she also wrote several commentaries on classical texts. It is thought that she assisted him on the commentaries to Euclid's *Elements* to which we referred earlier. Of special importance was her commentary on Diophantus's thirteen-volume *Arithmetica*, which had been written circa 250, and an eight-volume popularization ("popular," it must be admitted, is a relative term here) of Apollonius of Perga's *Treatise on Conic Sections*, which has not survived. What little we do have of the *Conic Sections* in Greek (half a book) may in fact be from this edition by Hypatia; the remainder only survived the Middle Ages in Arabic translations. But it was a book that greatly influenced the work of scientists such as Kepler, Newton, and Descartes (Apollonius's conic sections were, in fact, decisive for Kepler's work on planetary orbits).

---

[7] Synesius had gone to Athens to continue his studies in philosophy, but he was tremendously disappointed. According to his account, the Athenian philosophers did not reach the sole of Hypatia's shoes: "Today's Athens has nothing venerable but the famous names of the places [. . .] Undoubtedly, today, in our time it is Egypt that has received and germinates the seed of Hypatia."

In her correspondence with Synesius, there are also several diagrams and descriptions which show that Hypatia designed several scientific instruments, proving that she was not a purely theoretical scientist. She and Synesius together designed an astrolabe that he then had built and sent as a gift to a Roman officer named Paeonius[8]. In Synesius's letter accompanying the gift (for some, this is one of the most important documents in the history of astronomy) we read[9]:

> I am therefore offering you a gift most befitting for me to give, and for you to receive. It is a work of my own devising, including all that she, my most reverend teacher [Hypatia], helped to contribute, and it was executed by the best hand to be found in our country in the art of the silversmiths. [. . .] We worked it out and elaborated a treatise and studded it thickly with the necessary abundance and variety of theorems. Then we made haste to translate our conclusions into a material form, and finally executed a fairest image of the cosmic advance.

In another letter, Synesius, gravely afflicted with the illness that would end his life, asks Hypatia to build an instrument to measure liquid densities, for medical purposes: "I am in such evil fortune that I need a hydroscope. See that one is cast in brass for me and put together" (today, the invention of the hydrometer is attributed to Hypatia). It remains a bit of a mystery to what use Synesius intended to put the hydrometer, but ancient medicine was not quite as advanced as ancient astronomy.

As Hypatia's academic work—both research and teaching—flourished, life in Alexandria went on under a tense and unstable peace, until the year 380, when Emperor Theodosius I, the first emperor not to take the title of *Pontifex Maximus*, made Christianity the official religion of the Roman Empire. Theodosius not only ended support for traditional Roman religion, he forbade the public worship of the ancient gods in the first systemic effort to end paganism. With such support from the State, the power of the Church over Alexandrian public affairs began to grow. The accession of Theophilus as bishop of Alexandria in the year 385 marked this turning point for Alexandria. Theophilus was a cultured and well-connected man and a friend of Synesius, with notable influence among the upper classes of Alexandria, who nonetheless persecuted

---

[8] Alan Cameron and Jacqueline Long (1993). *Barbarians and politics at the court of Arcadius.* University of California Press.

[9] Livius.org. Articles on ancient history. http://www.livius.org/sources/content/synesius/synesius-on-an-astrolabe/synesius-on-an-astrolabe-3/

paganism with an iron hand. But the pagan population of Alexandria, the supporters of Greco-Roman culture, belonged to the upper classes for the most part and formed a powerful pressure group. Theophilus could not fight them without a good excuse to change the situation.

This arrived in the year 391, when the emperor issued a new decree against the pagan world stating that "no one will go to the sanctuaries, walk through the temples, or raise their eyes to statues created by the work of man," a move that, as so often in the history of religions, forced its targets underground, with the appearance of hidden temples inside private homes. One of these temples was discovered that same year by the bishop of Alexandria, who, along with his followers, looted it, publicly exposed its idols, and made fun of them. This offense unleashed the wrath of pagan believers, who attacked Christians. But the Christian believers fought back, and, with greater numbers, forced their withdrawal into the Serapeum. Other pagans came to their defense (fighting for their friends and/or their ideals), but they were either killed or ended up taking refuge in the temple. In the end, the pagan contingent was trapped inside the Serapeum, along with numerous academics (our sources are silent on whether or not Hypatia was among them), under a state of siege.

Theophilus brokered a deal: he had the emperor send him a letter authorizing him to destroy the Serapeum and to proclaim as martyrs those Christians who had died, in exchange for pardoning the pagans trapped inside. Under the watchful eye of the army, they were liberated and the temple demolished. In its place, a Christian church was built that existed until the tenth century. The destruction of the Serapeum, and most probably of its collection of books (although it is not unlikely that the cultured Theophilus augmented his personal collection with a good number of them), marked the end of the institutions of the Museum and the Library of Alexandria, some seven hundred years after the Museum's founding. Carl Sagan said of this event[10]: "It was as if the entire civilization had undergone some self-inflicted brain surgery and most of its memories, discoveries, ideas, and passions were extinguished irrevocably."

The following years were convulsive. Theodosius I died in 395 and the Roman Empire was divided into halves ruled by his sons, the western part ruled by Honorius in Rome and the eastern (what we now

---

[10] Carl Sagan and Ann Druyan (2013). *Cosmos*, p. 356. Ballantine.

know as the Byzantine Empire) ruled by his brother Arcadius from Constantinople, including Egypt. Under Arcadius, the pagans suffered more pressure, and many converted to Christianity, although Hypatia did not. In 412, Bishop Theophilus died, and his nephew Cyril took his place (albeit not peacefully and with strong opposition). That same year, Orestes, another of her former pupils—and recently baptized in Constantinople—was appointed as governor of Egypt, the highest regional representative of the Emperor of the Eastern Empire. Hypatia, meanwhile, continued to operate her private academy and also to pursue an active role in the political life of Alexandria. Through former students (although she was also consulted by those unaffiliated with the school), she had a remarkable authority and influence and was frequently consulted by the magistrates about matters concerning the administration of the city. As soon as Orestes arrived in Alexandria as governor, he and Hypatia refreshed their relationship. Hypatia became a frequent visitor to the governor, for Orestes routinely asked for her advice. In turn, he warned her that she was setting herself up as an adversary to Bishop Cyril and advised her on numerous occasions to convert to Christianity, which she did not.

Many things can be said about Bishop Cyril (c.376–444) except that he was a man of easy character. His friction with Roman authority was continuous, and he seemed to seek out conflict in every direction. He renewed the persecution of paganism, and extended it to Christian heretics and Jews. He forcibly converted numerous synagogues into churches, through all these actions fomenting unrest in Alexandria. These disturbances of the delicate religious and political order were finally denounced by Orestes to the emperor in 414, as Cyril was chipping away at the emperor's power. The bishop took this denunciation as a threat to his own power in turn, and he and his allies responded with violence: several hundred Nitrian desert monks came to his aid, and, led by a certain monk named Ammonius, they attacked Orestes when he was in a carriage and left him badly wounded. The situation escalated from there: Orestes had Ammonius tortured and executed, and in return, Cyril honored Ammonius as a martyr and made his break with the imperial representative final.

That year 414 would only become worse for Hypatia. In addition to the attack on Orestes, who showed her how insecure her situation was, her other powerful friend, Bishop Synesius, died. Perhaps if he had remained alive—given his relationship with Cyril's uncle, Theophilus—things might have gone differently. But Hypatia no longer had even this shield, and Cyril felt a marked antipathy towards her, whom he blamed for having turned Orestes against him. Her power and influence in Alexandria—compounded by the fact that she was a woman!—only made things worse. Finally, he accused her publicly of witchcraft, and of having put the governor under her spell. This inflamed a Christian mob, which on a terrible day in the spring of 415 while Hypatia was returning home in a carriage, attacked her, pulled her from the carriage, undressed her, and skinned her alive. With her death, Cyril was cheered by the masses as a savior, as a "new Theophilus" who had eliminated from Alexandria any remnant of paganism. Orestes asked the emperor for an investigation into this murder but, in a clear sign of his diminished political power, the request was dismissed for lack of witnesses. Without the support of Hypatia and with the direct opposition of the Church, Orestes resigned his position as governor and left Alexandria[11].

We know the story of this grisly event almost first hand. Several Alexandrians fled the city after these events and took refuge in Constantinople. There they told the story of what happened to a Christian Greek historian named Socrates Scholasticus (also known as Socrates of Constantinople) who recorded it in his *Historia Ecclesiastica*. Socrates, though a Christian, clearly admired Hypatia and pointed to Cyril as the main architect of her murder. We opened this chapter with Socrates's summary of Hypatia's achievements; his account continues:

> On account of the self-possession and ease of manner, which she had acquired in consequence of the cultivation of her mind, she not infrequently appeared in public in presence of the magistrates. Neither did she feel abashed in coming to an assembly of men. For all men on account of her extraordinary dignity and virtue admired her the more. Yet even she fell a victim to the political jealousy which at that time prevailed. For as she had frequent interviews with Orestes, it was calumniously reported among the Christian populace, that it was she who prevented Orestes from being reconciled to the bishop. Some of them, therefore, hurried away by a fierce and bigoted zeal, whose ringleader was a reader named Peter, waylaid her returning home, and dragging her from her carriage, they took her to the church called Cæsareum,

---

[11] He made the right decision. The governor who succeeded him, Calisto, was killed a few years later by another mob.

where they completely stripped her, and then murdered her with tiles. After tearing her body in pieces, they took her mangled limbs to a place called Cinaron, and there burnt them. This affair brought not the least opprobrium, not only upon Cyril but also upon the whole Alexandrian church. And surely nothing can be farther from the spirit of Christianity than the allowance of massacres, fights, and transactions of that sort. This happened in the month of March during Lent, in the fourth year of Cyril's episcopate, under the tenth consulate of Honorius, and the sixth of Theodosius.

The Alexandrian Christians must have had a guilty conscience for this terrible deed: there is an alleged letter from Hypatia to Cyril, a forgery written shortly after her death, in which she "confesses" to professing the Nestorian heresy[12]. By the standards of religious conflict of the day, killing a heretic of your own religion was more justifiable than extramural religious murder. The death of Hypatia was a tragic event, marking the end of the ancient world and the beginning of the dominance of the Christian religion—and a prolonged stagnation of scientific progress. The shreds of knowledge that were saved from the destruction of the Library slumbered for a long time until the Arab world rediscovered them several centuries later.

Hypatia was all but forgotten by history; Cyril was made a saint.

With the Enlightenment, Hypatia's posthumous fortunes turned, and she joined the pantheon of scientific martyrs, a victim of irrationality along with Anaxagoras, Giordano Bruno, Miguel Servet, and Antoine Lavoisier (among many others). In 1651, Riccioli baptized a crater with the name of Hypatia, in the region that we might call "Alexandrian" (because of the many Alexandrine characters that he located in this region). In 1884, the astronomer Víctor Knorr named an asteroid with her name. In 1973, the IAU moved the name of Hypatia to another, smaller crater southwest of the Sea of Tranquility. Crater Hypatia is elongated, approximately 25×17 miles, probably because the impact that formed it was very oblique. Near to it, to the south, are the craters Theophilus and Cyrillus, named in honor of the two bishops of Alexandria, uncle and nephew, and the crater for Catharina, another

---

[12] This dogma was defended by Nestorius, bishop of Constantinople, and contrary to the beliefs of Cyril himself. It proposes that Jesus had two different natures, being half human, half divine (dyophysitism) while the mainstream accepted that both natures existed but were joined in one same essence (hypostasis).

Alexandrian whom we will discuss in the next chapter. To the north of Hypatia, in the middle of the Sea of Tranquility, there is also a set of faults in the lunar terrain (or rilles; in Latin: *rimae*, to use the geological term) called Rimae Hypatia. This is, by happy historical accident, very close to where Neil Armstrong (1930–2012) stepped on the Moon for the first time, followed by his partner Buzz Aldrin (born in 1930) at Tranquility Base[13].

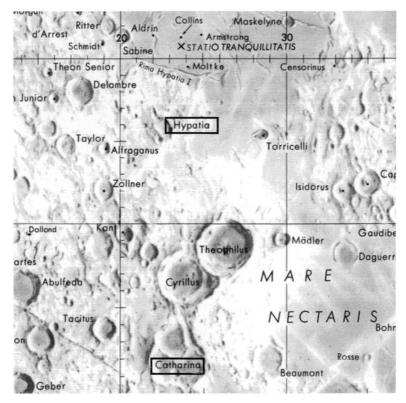

Figure 18 Location of craters Hypatia and Catharina. Courtesy of the Lunar and Planetary Institute, Houston, Texas.

[13] For those craters which are on the visible part of the Moon we provide maps courtesy of the Lunar and Planetary Institute, Houston, Texas, should you wish to find them with a telescope. For those on the far side, we show a smaller map and images from NASA's Lunar Reconnaissance Orbiter (LRO—http://lroc.sese.asu.edu) NASA/GSFC/ Arizona State University, or other sources as noted.

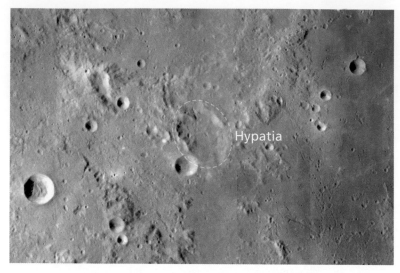

Figure 19 Lunar Reconnaissance Orbiter zoom on crater Hypatia (image width is 150 miles).

2

Catherine of Alexandria (*c*.287–*c*.305)

Figure 20 Saint Catherine of Alexandria (Carlo Crivelli *c*.1470).

St. Catherine, glorious virgin and martyr, resplendent in the luster of wisdom and purity; thy wisdom refuted the adversaries of divine truth and covered them with confusion; thy immaculate purity made thee a spouse of Christ, so that after thy glorious martyrdom angels carried thy body to Mount Sinai. Implore for my progress in the science of the saints and the virtue of holy purity, that vanquishing the enemies of my soul, I may be victorious in my last combat and after death be conducted by the angels into the eternal beatitude of heaven. Amen.

Saint Catherine of Alexandria was a Christian martyr who, according to tradition, lived between 287 and 305. After the Virgin Mary, she is possibly the most widely represented woman in Christian iconography. Despite this, there is almost no historical evidence for her life—or even existence; in the words of two recent scholars[1]: "Such in very brief outline is the tale which was the starting point of the cults of St. Catherine, of which the ramifications were many and interesting in both East and West. She is in the Roman Martyrology, but there is no other positive evidence that she ever existed outside the mind of some Greek writer who first composed what he intended to be simply an edifying romance. Her emblem in art is a wheel."

---

[1] Donald Attwater and Catherine Rachel John (1996). *The Penguin Dictionary of Saints*, third edition. Penguin.

The Women of the Moon. Daniel R. Altschuler Stern and Fernando J. Ballesteros Roselló.
© Daniel R. Altschuler Stern and Fernando J. Ballesteros Roselló 2019. Published in 2019 by Oxford University Press. DOI: 10.1093/oso/9780198844419.001.0001

How she came to have a memorial on the moon is a testament to the power of story. Her biography (insofar as it is not clearly folktale) is brief. Tradition makes her the offspring of a noble family, daughter of the Sicilian king Costus and his queen Sabinela, Roman imperial governors of Alexandria. Unfortunately, this royal couple does not appear anywhere in historical records except in those texts that speak precisely of Catherine's life. Considering, then, that we are largely in the world of tradition rather than history, the story goes that although hers was a pagan family, at one point during her adolescence, Catherine received a mystical visit from the Virgin Mary, who introduced her to her son, Jesus, and made her an ardent convert to Christianity.

All accounts of Catherine's life and martyrdom (we'll get to that soon) were written well after her death, with the earliest dating from the seventh century; these early tales are much less detailed than those that would follow over the rest of the Middle Ages. The Catholic Church removed her from its list of saints in 1969 because of the great doubts about her historical existence[2]. The main reservations about her existence spring not just from the lateness of the biographical information we have, but from the substance of her biography itself, which is very reminiscent of that of Hypatia, a near contemporary. Both women came from noble and well-situated families. They had a reputation for great beauty and were renowned for their precocious wisdom, vast knowledge, mastery of philosophy, arts, and sciences, and their use of reason and dialectics. They rejected marriage on all occasions from even the most well-placed men. The Roman authorities ordered Catherine to renounce her faith, which she refused to do, and so she was tortured and skinned alive (using a wheel with sharp knives). Although she miraculously recovered from this event (the wheel fell apart), she was finally beheaded. If we remove the miraculous recoveries and replace "not wanting to renounce Christianity" for "not wanting to accept Christianity," the parallelism between the biographies of both women is remarkable.

For many historians, St. Catherine and Hypatia are really the same person. Catherine, in this understanding, is an invention of the late antique Church, created in opposition to the figure of the great philosopher of Alexandria. The great tendency to syncretism that

---

[2] However, in 2002 the Church was forced to reincorporate her martyrdom, due to popular devotion.

primitive Christianity had is well known: the Virgin Mary, represented on many occasions with a moon at her feet, was identified with several Roman goddesses, among them Diana—goddess of the moon—and Maia[3] (similar in name to the Virgin), whose celebratory month of May would become dedicated to Mary in the Christian calendar. Likewise, even Jesus himself was identified with the god Apollo, the solar god, because, like Apollo, he dies at night and returns during the day (and thus the first Christian religious buildings had their entrance door to the east, towards the rising sun, where the god was resurrected). These are just a few examples, but similar cases abound, so it seems quite likely that a story as well known as Hypatia's was Christianized, integrating it into the mythology/history of nascent Christianity (and, we venture to point out, in the process recasting a horrific event that did not place the Alexandrian Church in a very good light).

Of course, Catherine could not be directly contemporary to the historical Hypatia; she had to be placed at a previous time, since, in the year 313, Emperor Constantine I had legalized Christianity. From that moment the Christian religion would gain ground, and in the year 380 the Emperor Theodosius I made it the Empire's official religion. It is clear that if Catherine had died because she did not renounce Christianity, she could not have lived at this time. The tradition, therefore, would have placed her death at an earlier time: in 305, eight years before the edict of Constantine I.

Whether the history of Catherine is historically true or not, she was one of the most influential figures in the history of the Church (the same might be said of God). Tradition has it that her remains were moved by angels to a monastery located at the foot of Mount Sinai. This monastery, belonging to the Orthodox Church of Jerusalem, is one of the oldest in Christendom. It was founded in the sixth century in the place where God, it is said, appeared before Moses in the burning bush; in fact, in the monastery, an old bush is preserved, said to be the protagonist of that miracle. The monastery of Mount Sinai has been inhabited continuously since its inception. Around the year 800, its monks found in a cave on Mount Sinai some female remains that they identified without hesitation as those of Saint Catherine. In a somewhat rare piece of demonstrable fact

---

[3] The goddess Maia, also called Bona Dea (the good goddess), was a virgin goddess whose cult was linked to virginity (of great importance in the Greco-Roman world); she was patron of female fertility and motherhood. It is no coincidence that even today, Mother's Day is celebrated in May.

in this narrative, someone's bones were found, and these remains of a woman were taken to the monastery, where they are now buried. Since no relative of Catherine attended to the removal of the corpse, there was no one to deny the identification that the monks made.

From that moment, the monastery became one of the main centers of Christian pilgrimage, and received the name of Saint Catherine's Monastery (while it is officially known as the "Sacred Monastery of the God-Trodden Mount Sinai"). During the Crusades, the legend of the saint was discovered by the medieval knights, who spread it throughout Europe, giving rise to great devotion. Since the tradition assumes Catherine was versed in dialectics, sciences, arts, and philosophy, she is the patron of philosophers and university students (she is more specifically the patron of the universities of Paris and Oviedo, and innumerable schools below the university level) and promoter of knowledge in general.

For her patronage of the sciences, Riccioli dedicated crater Catharina, located northwest of the Sea of Nectar, to her, and if she and Hypatia are indeed the same person, it would be the second lunar crater dedicated to the Alexandrian astronomer. Catharina is an ancient crater, a circular formation highly eroded by meteoric impacts. At about 100 km in diameter, it is one of the largest dedicated to a woman. You can see some interesting features in its northeast face and it has smaller craters inside. Apollo 16 landed about 300 km from it. As a great irony, the crater is next to two other large craters: Cyrillus, dedicated to St. Cyril of Alexandria, the instigator of Hypatia's death, and Theophilus, Cyril's uncle and also bishop of Alexandria. Together, these three craters form a set easily recognized even with binoculars (see Figure 18 in the previous chapter and Figure 21).

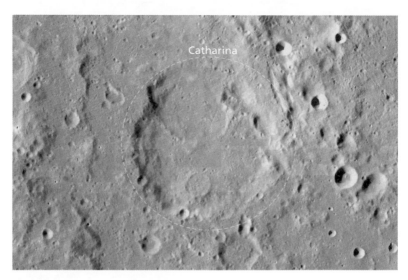

Figure 21 Lunar Reconnaissance Orbiter zoom on crater Catharina (image width is 150 miles).

3

Nicole-Reine de la Briere Lepaute (1723–1788)

Of sine tables always surrounded,
You followed Hipparchus and Ptolemy
But it is not enough to follow in their
    footsteps,
And to be amongst those that we fill with
    honor,
Queen, if you were not the sine of the
    Graces,
And the tangent of our hearts
        A poem dedicated to Nicole-Reine
        Lepaute by the French astronomer
                    JOSEPH LALANDE.

Figure 22 Nicole-Reine de la Brière Lepaute. Bibliothèque Nationale de France.

Nicole-Reine Lepaute was a leading astronomer of prerevolutionary France. During a time of great inequality—social, economic, and political—Lepaute was fortunate to have been born into advantage, and she made the most of her opportunities. While few could vote, and the rigid social system dominated by the nobility and the clergy made clear to everyone what their place was in the world, the ignorance in which the populace was kept stood in marked contrast to the enlightened despots ruling Europe, promoters of the scientific academies that would provide the intellectual impulse of the Enlightenment.

Thanks to this impulse, a plethora of scientists thrived (although post-revolutionary France was not short on towering intellects, it must be noted) such as Lavoisier, Legendre, Laplace, Lagrange . . . Almost all of them belonged to well-situated families or to the nobility, as shown by the great abundance of surnames beginning with La, Le, and L' (after

The Women of the Moon. Daniel R. Altschuler Stern and Fernando J. Ballesteros Roselló.
© Daniel R. Altschuler Stern and Fernando J. Ballesteros Roselló 2019. Published in 2019 by Oxford University Press. DOI: 10.1093/oso/9780198844419.001.0001

the Revolution, many of them merged the article with what came next to disguise their social status). Science was held in high regard among the upper layers of society. Rulers liked to be surrounded by intellectuals as a symbol of distinction, aware of the role that mathematics and science had to drive industry and military power. The title of Member of the Académie des Sciences was, in fact, as dear as a title of nobility. This esteem for science was paired with notions of gender, such that intellectualism was seen to enhance the life and character of a woman, and since knowledge of accounting was useful for the wife of a powerful man, mathematics (and related disciplines such as physics or astronomy) came to be suitable subjects for women. Many women of high society took advantage of this ephemeral opportunity and acquired a scientific background, among them Nicole-Reine Lepaute.

Nicole-Reine Etable was born in 1723, virtually surrounded by royalty at the Luxembourg Palace in Paris. Her father, Jean Etable, was a valet in the service of Louise Élisabeth d'Orléans, daughter of Philippe II, Duke of Orléans (regent of France during the childhood of Louis XV)[1]. Largely self-taught (like almost all the women of the Moon) through books and periodicals, Nicole showed a marked intellectual and social brilliance from a very young age. She lived all her life in the Luxembourg Palace with her husband, the royal watchmaker Jean-André Lepaute, whom she met when he went to the palace to install a new clock. She married him at twenty-five. The work of watchmaker was much more closely linked to astronomy at that time than today; after all, clocks must accurately reproduce the Earth's rotation, and high-precision clocks were essential for both astronomers and navigators. Thanks in part to interaction with her husband's work, Lepaute became interested in astronomy and mathematics: she began to perform astronomical observations in order to improve the accuracy of his clocks, and she also performed the tedious mathematical calculations associated with these observations. Over time, she attained mastery of both astronomical observation and mathematical developments, her fame in these

---

[1] Poor Louise Élisabeth had been married the previous year, aged only twelve, to Luis, the Prince of Asturias and heir to the throne of Spain, aged fourteen. The following year, after the abdication of the Spanish King, Felipe V, in favor of his son Luis, Louise Élisabeth would become Queen of Spain, during the brief span of eight months. If that breakneck career pace were not enough, Luis I of Spain died of smallpox a few months after having acceded to the throne, and soon after, Louise Élisabeth retired to France and the Luxembourg Palace, where she would spend the rest of her life.

tasks grew, and she came to acquire among her acquaintances the nickname of *savante calculatrice*, the "savant calculator" (or in a more "modern" translation, the "scientific calculator").

Shortly after their wedding, a young nobleman settled in the
Luxembourg Palace; although a law student, he was fascinated by
astronomy, and he was given a room with a porch that he could use as
an observatory. He was Jérôme Lefrançais de Lalande (1732–1807, later
known as Jérôme Lalande) with whom the young couple would establish a lifelong friendship. In 1750, Lalande determined that he had
had his fill of palatial observations and that he would devote his life
to science (once he finished his law studies). He soon joined the 1751
expedition of Nicolas-Louis de Lacaille (1713–1762), one of the eighteenth
century's greatest observational astronomers, to the Cape of Good Hope,
to carry out a series of lunar parallax observations in order to correctly
measure the distance between the Earth and the Moon. Their observations from Berlin and the Cape yielded a distance of 226,800 miles, only
about five percent off from the true average distance of 238,856 miles[2].
The success of this task opened for Lalande, on his return, the doors of
the Académie des Sciences, which he joined as an assistant astronomer.

Lepaute, meanwhile, and her husband designed and built an astronomical pendulum clock for astronomers of the Académie (it happened that Lalande was in charge of testing the instrument; he gave his
approval). The operation of a normal pendulum clock is extremely sus-
ceptible to changes in temperature: expansion and contraction of the
pendulum results in the clock running alternately faster (in cold conditions) and slower (in hot environments). After a short time, the accumulated error can be so large that it becomes unacceptable for an
astronomer. Lepaute became an expert on mechanisms to counteract
this source of inaccuracy, and among her writings is the "Table des lon-
gueurs des pendules" (Table of the lengths of pendulums), included in
her husband's *Traité d'Horlogerie* (Treatise on Watchmaking) of 1755.

For his part, Lalande wanted to dazzle France by calculating precisely
the time of the next passage of Halley's Comet. He posed the question
to his colleague Alexis Clairaut (1713–1765), who told him that the main
problem was to calculate the perturbations that different planets produced in the comet's orbit. Clairaut proposed to Lalande an ambitious

---

[2] Due to its elliptical shape, the lunar distance varies from 356,500 km (221,500 mi) to
406,700 km (252,700 mi), with an average value of 384,402 km (238,856 mi).

program of long and tedious calculations with which the problem could be solved. Given the size of the task, Lalande decided to include in the team the person he considered most appropriate for mathematical calculation, his friend the *savante calculatrice*: conscientious, systematic, and error-proof. Lepaute accepted the proposal and the three locked themselves up for six months in 1758 to perform the calculations that would allow them to predict when Halley's Comet would return. Lalande would write of the time in *Bibliographie Astronomique* (1803):

> During six months we calculated from morning to night, sometimes even at meals, the consequence of which was that I contracted a disease which changed my constitution for life. The assistance of Mme Lepaute was such that, without her I should never have been able to undertake the enormous labor, in which it was necessary to calculate the distance of each of the two planets Jupiter and Saturn from the comet, separately for each successive degree for 150 years.

The result was a successful prediction, considering that the prolonged calculation had been done by hand: the comet, they said, would pass again by its closest point to the Sun (perihelion) in mid April 1759, with a margin of error of one month. And indeed, the comet reached its perihelion on March 13, just within the margin of error. After that, Clairaut published his *Théorie des cometes* (Theory of Comets), omitting all mention of Lepaute and the work she had done. The cause of this exclusion was Clairaut's fiancée, one Miss Marie-Anne Goulier, who was jealous of Lepaute (perhaps for spending six months locked up with her boyfriend). But in return, the omission provoked Lalande's anger and precipitated a permanent break with Clairaut, with whom he never worked again.

Instead, Lalande continued to collaborate frequently with Nicole-Reine Lepaute. From 1760 to 1776, Lalande directed the magazine of the Académie des Sciences, and Lepaute was in charge of calculating the tables of astronomical ephemerides for this publication. She also made and published the calculations for the annular eclipse of 1764, and drew a map of the places and times from which it would be visible that was widely distributed throughout Europe. But the milestone that marked her for posterity (albeit indirectly) was the transit of Venus across the Sun in 1761. Transits were especially important to astronomers because measuring the time of the transit from different places on Earth allows the distances between the Earth, Sun, and Venus to be calculated with great precision. For the magazine of the Académie, Lepaute calculated with Lalande the ephemerides of the phenomenon, and later published a supplement with the observations of the transit of Venus. Lepaute's work was formally recognized that year as she was accepted as a member of the prestigious Académie de Béziers and a flower was dedicated to her: the hydrangea (*Hortensia*). To understand how "Hortensia" is named after Nicole-Reine Lepaute we need to include another character in the narrative, the most hapless astronomer in history: Le Gentil.

Guillaume Le Gentil (Guillaume Joseph Hyacinthe Jean-Baptiste Le Gentil de la Galaisière, 1725–1792) sailed to India to observe the 1761 transit of Venus from the city of Pondicherry (nowadays Puducherry), on the eastern coast of the country, then part of the French colonial empire. While he was at sea, however, war broke out between France and England over the colonial rule of India, and on arrival he found that the city had been taken by the English army. To avoid being taken prisoner—although there was little time left until the day of transit—he and the crew decided to go to the island of Mauritius in the Indian Ocean and observe from there. Unfortunately, they did not arrive in time, and the transit took place while they were still at sea. Le Gentil tried to do the observations from the ship, but the moving deck made it impossible.

Fortunately, transits of Venus occur in pairs separated by eight years[3]; in 1769 he would have a second chance. Le Gentil chose to stay those eight years, and devoted himself to studying the people and nature of several Indian Ocean islands, including Madagascar. Pondicherry soon returned to French hands. So, on March 27, 1768, after some dangerous travels and with time to prepare, he returned to Pondicherry. There he built a small observatory to measure the transit accurately. In the meantime, he explored the area and discovered several botanical species unknown in Europe, including a beautiful flower that grew in close clusters and from which he took various seeds. At last the day of the transit arrived, but unexpectedly (because for a month the weather had been excellent) the day began totally covered with clouds. In his words: "At three or four minutes before seven o'clock, almost the moment when Venus was to go off the sun, a light whiteness was seen in the sky which gave a suspicion of the position of the sun, nothing could be distinguished by the telescope." The observation was impossible; the

---

[3] But these pairs are separated from the next pair by more than a century.

expedition had been a failure. In his memoirs he writes[4]: "That is the fate which often awaits astronomers. I had gone more than ten thousand leagues; it seemed that I had crossed such a great expanse of seas, exiling myself from my native land, only to be the spectator of a fatal cloud which came to place itself before the sun at the precise moment of my observation, to carry off from me the fruits of my pains and of my fatigues." Completely discouraged and after nine years absent from home, he undertook the return to France in 1770. His misfortunes had not yet ended. On the return trip he fell ill with dysentery and stayed to recover on the island of Reunion. There he had to wait yet another year for a Europe-bound vessel. Finally, after eleven years absent, he arrived in France . . . and found that he had been given up for dead, his wife had remarried, his property had been distributed among several heirs, and his place in the Académie was occupied by another astronomer.

After some years of litigation and mediation (and with the king's help) he finally got his life back (with a new wife). However, he had managed to bring with him the seeds of a flower he had found in India, and he decided to give the flower a name in honor, precisely, of Nicole-Reine Lepaute, who had earned the epithet (due to her rarity as an astronomer's astronomer, and her efficiency) a "flower of the gardens," or in Latin, *flos hortorum*. The flower was therefore dubbed "Hortensia." In fact, in her life Lepaute was so identified with the flower that many ended up believing that, indeed, that was her real first name. Her own nephews called her "Aunt Hortensia," and she appears in several encyclopedias under that name. Even the National Library of France preserves a portrait of Nicole-Reine Lepaute inscribed with the name "Hortense Lepaute."

She lived a happy life with her husband, and the two died just a few months apart in 1788, a year before the French Revolution. Her Moon crater was baptized in 1898 by the Bavarian selenographer Johann Krieger. This small crater, 16 km in diameter, is located on the visible side of the Moon, towards the southwest near the lunar edge, in the Palus Epidemiarum ("The Swampland[5] of the Epidemics"). An asteroid, too, bears her name: asteroid 7720, discovered in 1960 in the asteroid belt.

---

[4] Helen Sawyer Hogg (1951). Out of Old Books (Le Gentil and the Transits of Venus, 1761 and 1769). *Journal of the Royal Astronomical Society of Canada*, Vol. 45, p. 142.

[5] *Palus* may be recognized in "paludism," an archaic name for malaria. The Italian for "bad air" caught more traction than the Latin for "swampery," evidently.

Figure 23 Location map for crater Lepaute. Courtesy of the Lunar and Planetary Institute, Houston, Texas.

Figure 24 Lunar Reconnaissance Orbiter zoom on crater Lepaute (image width is 150 miles).

4

Caroline Lucretia Herschel (1750–1848)

Figure 25 Caroline Herschel.
ETH-Bibliothek.

What is our age, if that age was dark? As for my name, it will also be forgotten, but I am not accused of being a sorceress, like Aganice, and the Christians do not threaten to drag me to church, to murder me, like they did Hypatia of Alexandria, the eloquent, young woman who devised the instruments used to accurately measure the position and motion of heavenly bodies. However long we live, life is short, so I work. And however important man becomes, he is nothing compared to the stars. There are secrets, dear sister, and it is for us to reveal them.

CAROLINE to her sister

Caroline Herschel's personal origins did not hint at her future scientific prowess[1]. She was born in Hanover in 1750, one of ten children of whom four did not survive childhood (typical of the time), and the youngest daughter of Isaac Herschel, military musician by profession, and Anna Ilse Moritzen, an illiterate peasant. These times were marked by the Seven Years' War, a conflagration that drew in much of Europe, with France, Austria, and Russia on the one side, and England and Prussia on the other. Between 1756 and 1763, one and a half million people were killed. The principality of Hanover was at that time part of the English kingdom, although its residents spoke German.

[1] Marilyn Ogilvie (2008). *Searching the stars: the story of Caroline Herschel*. The History Press. Michael D. Lemonick (2009). *The Georgian star*. W.W. Norton.

The Women of the Moon. Daniel R. Altschuler Stern and Fernando J. Ballesteros Roselló.
© Daniel R. Altschuler Stern and Fernando J. Ballesteros Roselló 2019. Published in 2019
by Oxford University Press. DOI: 10.1093/oso/9780198844419.001.0001

Isaac Herschel, although often absent on military service, supported the education of his children who learned the rudiments of reading, writing, and arithmetic, but Anna (like her contemporaries) thought that this was not for women and intended that their youngest daughter, Caroline, should become the servant of the house. Medical issues in childhood reinforced the low expectations her family had of her: smallpox marked her face when she was three years old, and at ten she contracted typhoid, stunting her growth; she would reach a height of only one and a half meters (four feet three inches) as an adult. Her father told her to give up hope of finding a husband because she was ugly and was not rich, and in fact, she never married.

Two of her brothers, Jacob (the eldest) and William (Friedrich Wilhelm, the second eldest), learned music from their father and enlisted in the military band, accompanying the Prussian armies to battle. After the French triumph at the Battle of Hastenbeck in 1757, while the French occupied Hanover, the brothers decided to emigrate, and undertook the two-week journey from Hamburg to London. In England, William began to work as a composer, performing musician (playing the violin, cello, piano, and organ), and music teacher. After several years working in different places, he obtained regular employment as the organist and musical director in the octagonal chapel of Bath in 1767, a city of thermal baths (hence its name), and at that time, an important cultural center for the English aristocracy.

He recruited his brothers Dietrich (the youngest), Alexander, and Jacob to play in his orchestra, and in 1772, William traveled to Hanover to rescue Caroline, then twenty-two years old, from her mother's mistreatment and took her to live in his house in Bath to be his housekeeper. In return, he had to hire a maid for his mother. Caroline, fearful and speaking no English, embarked on a journey whose final destination she could not imagine. Under the tutelage of her brother, Caroline learned to sing and began a successful career as a soprano singing in Händel's oratorios, which were very popular at that time.

William, meanwhile, was an avid reader and when he came upon the work of Robert Smith (1689–1768), who was a professor of Astronomy at the University of Cambridge and wrote about optics and telescope design, this sparked a transformation from musician to astronomer. William also read the popular astronomy book by James Ferguson (1710–1776), who argued that if God had created the stars it was because they gave light and gave heat to the inhabitants of their planets. He started to make his own telescopes because he did not have enough money to buy one, and his skills grew to the point where he was said to make the best telescopes in the world (he made dozens of telescopes for sale, and among his customers were some important observatories, including the Real Observatorio de Madrid. For them, he constructed a sixty cm reflector, which was destroyed by Napoleon's troops in 1808). The most difficult part of telescope making in Herschel's day was polishing the reflecting mirrors, made of an alloy of copper and tin. He first had to melt the metal in a specially constructed oven, and then cast it in a mold prepared with horse dung (a material abundant at that time when transportation was by horsepower). Glass mirrors covered with metal, lighter and easier to polish, were invented later, around 1856.

In March 1781, William was observing the sky in search of binary stars with his sixteen cm reflector (for that time, a good-sized telescope) in the constellation Gemini, when he spotted an object whose disk indicated that it was not a star. Successive observations showed that the object moved. He communicated this discovery, thinking that it was a peculiar comet, to his friend Dr. William Watson, member of the Royal Society, who in turn shared it with the royal astronomer Nevil Maskelyne (1732–1811). The Royal Society of London for Improving Natural Knowledge, founded in 1660 under the motto: *Nullius in verba* ("on the word of no one"), is one of the oldest extant scientific societies. One of its most important functions was to resolve conflicts over discovery priori—who invented or discovered it first?—and it also served as a conduit for the free exchange of ideas and results. Its motto is a clear rejection of sacred or philosophical books as a source of knowledge, and instead an implied admonition to read the "book of nature," as Galileo had proposed. The peer-review process was introduced by the first secretary and founding editor of the *Philosophical Transactions of the Royal Society*, Heinrich Oldenburg (1619–1677), a natural philosopher and German theologian[2].

Other astronomers, including the Frenchman Charles Messier (1730–1817), observed and confirmed the discovery, and some began to speculate that it was a planet—if true, a spectacular finding. This was confirmed by the orbit calculated from the observations by the Russian Andrei Ivanovich Leksel (1740–1784). By the end of 1781, the Royal

---

[2] *Philosophical Transactions of the Royal Society* was first published in 1665 and is still published. It is the oldest scientific publication in the world.

Society awarded the Copley medal[3]—roughly analogous to the modern-day Nobel Prize—to William for this great discovery. Herschel named the planet Georgium Sidus ("star of George"), in homage to King George III (1738–1820), who in turn named him "astronomer to the king" (Maskelyne was "royal astronomer," an official position), with an annual pension of two hundred pounds (the royal astronomer received three hundred). Perhaps unsurprisingly, the international astronomical community did not accept this name and after a while it was agreed to continue with the mythological tradition of the other planets and name it Uranus: father of Saturn, grandfather of Jupiter, and great-grandfather of Mars (in a case of one scientific discovery commemorating another, the element uranium, discovered in 1789, was named in honor of the planet's discovery).

Fame and recognition led William to devote himself full-time to astronomy. William and Caroline had to move to the vicinity of Windsor palace to be closer to the king. Gone was the daily pursuit of music for both of them, and Caroline's incipient singing career was particularly thwarted.

William and Caroline formed a research team that pursued astronomy beyond the solar system. With the telescopes built by William, they discovered and cataloged thousands of new objects, binary stars and clusters of stars, and nebulae (among them, distant galaxies whose nature they still did not know). The work was intense and exhausting. During the night, William gazed through the telescope and dictated to his sister the description of what he was seeing and the telescope's position. She recorded the data, so that he could observe without interruption. At dawn and during the day, she prepared tables and calculated the positions in the sky of the observed objects.

In 1783, William gave Caroline a telescope for her particular use, which, among other things, she devoted to a careful search for comets, and in 1787 King George III granted Caroline, as William's assistant, an annual pension of fifty pounds. The disfigured and almost illiterate girl had transformed, against all expectations, into a professional astronomer.

Between 1786 and 1797, Caroline, using her telescope, discovered eight comets. It is common to read that she was the first woman to discover a comet, but this is not precisely true; in 1702, the astronomer Maria Margarethe Winkelmann-Kirch (1670–1720), wife of the German

---

[3] Out of 265 annual awardees since 1731, only one woman has received a Copley Medal: Dorothy Crowfoot Hodgkin (1910–1994), who won it in 1976.

astronomer Gottfried Kirch (1639–1710), had discovered one. But Caroline held the record for *most* comet discoveries by a woman until 1980, when she was overtaken by another Caroline: Caroline Shoemaker (born 1929), who by 2002 had discovered thirty-two comets, and over 800 asteroids.

Having dedicated and shared her professional and personal life with her brother for fifty years, the death of William in 1822 at eighty-four years of age[4] was a traumatic event for Caroline. She felt alone, as "a person who has nothing else to do in this world"[5], and made the impulsive decision to return to Hanover after a lifetime abroad. She returned to live in the house of her only surviving brother, Dietrich, along with his wife and daughter (a decision which she later regretted). Fortunately for her state of mind, her nephew John Herschel (1792–1871; the son of William), himself a renowned astronomer, asked her to review the catalog of 2500 nebulas and star clusters that she had prepared with William, a job she completed in 1825. This work was of such importance that in 1828 the Astronomical Society of London[6] granted her one of the three gold medals it awarded that year. Not only was Herschel the first woman to obtain it, until it was awarded to the American astronomer Vera Rubin (1928–2016) in 1996, she remained the only one. On her medal, next to an engraving of the large forty-foot telescope built by William and Caroline, appears the society motto: *Quicquid Nitet Notandum*, which means "Observe everything that shines."

Caroline's nephew John Herschel, as president of the society, delivered a speech to the assembled members reviewing the life and the work of the awardees: along with Caroline, Sir Thomas Brisbane (1773–1869) and James Dunlop (1793–1848). At the end of his presentation Mr. South, the society's vice president, took the floor, and delivered an eloquent summary of Caroline's work[7]:

> The labours of Miss HERSCHEL are so intimately connected with, and are generally, so dependent upon, those of her illustrious brother, that an investigation of the latter is absolutely necessary ere we can form the

---

[4] The period of Uranus is also eighty-four years; in an evocative coincidence, therefore, Uranus was in the same position in the sky at his birth and his death.

[5] Marilyn B. Ogilvie (2008). *Searching the stars: the story of Caroline Herschel*, p. 187. The History Press.

[6] The Astronomical Society of London, founded in 1820, became the Royal Astronomical Society in 1831. The Royal Society of London is a different society.

[7] *Memoirs of the Astronomical Society of London*, 1829, Vol. 3, p. 409.

most remote idea of the extent of the former: but when it is considered that Sir W. HERSCHEL's contributions to Astronomical Science occupy sixty-seven Memoirs, communicated from time to time to the Royal Society, and embrace a period of forty years, it will not be expected that I should enter into their discussion.

To the Philosophical Transactions I must refer you, and shall content myself with the hasty mention of some of her more immediate claims to the distinction now conferred. To deliver an eulogy (however deserved) upon his memory is not the purpose for which I am placed here. His first catalogue of new nebula and clusters of stars, amounting in number to one thousand, was made from observations with the 20-feet reflector, in the years 1783, -4, and -5. A second thousand was furnished by means of the same instrument in 1785, -6, -7, and -8; whilst the places of five hundred others were discovered between 1788 and 1802. But when we have thus enumerated the results obtained in the course of sweeps with this instrument, and taken into consideration the extent and variety of the other observations, which were at the same time in progress, a most important part yet remains untold. Who participated in his toils? Who braved with him the inclemency of the weather? Who shared his privations? A female.—Who was she? His sister. Miss HERSCHEL it was, who by night acted as his amanuensis it was whose pen conveyed to paper his observations as they issued from his lips; she it was who noted the right ascensions and polar distances of the objects observed; she it was who having passed the night near the instruments, took the rough manuscripts to her cottage at the dawn of day, and produced a fair copy of the night's work on the subsequent morning; she it was who planned the labour of each succeeding night; she it was who reduced every observation, and made every calculation; she it was who arranged everything in systematic order; and she it was who helped him to obtain an imperishable name.

But her claims to our gratitude end not here; as an original observer she demands, and I am sure she has, our most unfeigned thanks. Occasionally her immediate attendance during the observations could be dispensed with.

Did she pass the night in repose? No such thing; wherever her illustrious brother was, there you were sure to find her also. A sweeper planted on the lawn became her object of amusement, but her amusements were of the higher order, and to them we stand indebted for the discovery of the comet of 1786; of the comet of 1788; of the comet of 1791; of the comet of 1793; and of the comet of 1795, since rendered familiar to us by the remarkable discovery of ENCKE. Many also of the nebulae contained in Sir WILLIAM HERSCHEL's catalogues were detected by her during these hours of enjoyment. Indeed, in looking at the joint labours of these extraordinary personages, we scarcely know, whether most to admire the intellectual power of the brother, or the unconquerable industry of the sister.

In the year 1797, she presented to the Royal Society a catalogue of 560 stars taken from FLAMSTEED's observations, and not inserted in the British catalogue; together with a collection of errata that should be noticed in the same volume. Shortly after the death of her brother, Miss HERSCHEL returned to Hanover. Unwilling, however, to relinquish her astronomical labours whilst anything useful presented itself, she undertook, and completed the laborious reduction of the places of 2500 nebulae, to the 1st Jan. 1800, presenting in one view the results of all Sir WILLIAM HERSCHEL's observations on those bodies; thus, bringing to a close half a century spent in astronomical labour. For this more immediately, and to mark their estimation of services rendered during a whole life to Astronomy, your Council resolved to confer on her the distinction of a medal of this Society. The peculiarity of our President's situation, however, and the earnest manner in which the feelings, naturally arising from it, were urged when the subject was first brought forward, caused your Council to waive on that occasion the actual passing

For this more immediately, and to mark their estimation of services rendered during a whole life to Astronomy, your Council resolved to confer on her the distinction of a medal of this Society. The peculiarity of our President's situation, however, and the earnest manner in which the feelings, naturally arising from it, were urged when the subject was first brought forward, caused your Council to waive on that occasion the actual passing their proposed vote. The discussion was however renewed on Monday last; and although there was every disposition to meet the President's wishes, still under a conviction that the doing so would have been a dereliction of public duty, it was Resolved unanimously, that a Gold Medal of this Society be given to Miss CAROLINE HERSCHEL, for her recent reduction, to January 1800, of the Nebulæ discovered by her illustrious brother, which may be considered as the completion of a series of exertions probably unparalleled either in magnitude or importance, in the annals of astronomical labour vote which I am sure everyone whom I have the honour to address, will most heartily confirm.

Mr. HERSCHEL,—In the name Of the Astronomical Society of I present this Medal to your illustrious Aunt. In transmitting it to her, assure her that since the foundation of this Society no one has been adjudged, which has been earned by services such as hers. Convey to her our unfeigned regret that she is not resident amongst us; and join to it our wishes, nay, our prayers, that as her former days have been glorious, so her future may be happy.

In 1835, the council of the Royal Astronomical Society (RAS)[8] resolved that:

> ... submits, that while the tests of astronomical merit should in no case be applied to the works of a woman less severely than to those of a man, the sex of the former should no longer be an obstacle to her receiving any acknowledgment which might be held due to the latter. And your Council therefore recommends this meeting to add to the list of honorary members the names of Miss CAROLINE HERSCHEL and Mrs. SOMERVILLE, of whose astronomical knowledge, and of the utility of the ends to which it has been applied, it is not necessary to recount the proofs.

Despite this statement of equality in awards, it was not until 1915 that the society was opened to female membership.

John Herschel, it may be noted, used the new catalog of Caroline to reexamine and improve the observations of his father and his aunt, extending it to the southern hemisphere and adding 1700 objects as a result of observations made from Cape Town in South Africa. The catalog was published in 1864 as the *General Catalogue of Nebulae and Clusters of Stars*. In 1888, the catalog was expanded by the Danish-Irish astronomer John Dreyer (1852–1926) and published as the *New General Catalogue of Nebulae and Clusters of Stars* (known as NGC) with 7840 objects. Many in the present know the NGC number of some object (the great Andromeda galaxy is NGC 221 and the Crab Nebula is NGC 1952), but few know that it all started with William and Caroline.

Despite the wish expressed by Mr. South, Caroline's final years were unhappy. In a letter to her nephew John of 1832, with her brother Dietrich having already passed away, she says[9]: "of the last 10 years I have spent here, I can only say that they have been a perfect tissue of disgusting vexations doubly painful to bear because I could not communicate my complaints to anyone; because they were against my immediate relations."

In 1846, at the age of ninety-six, she was awarded a Gold Medal for Science by the King of Prussia, conveyed to her by Alexander von Humboldt. Caroline died at ninety-eight years of age in Hanover, on

---

[8] *Memoirs of the Royal Astronomical Society*, 1835, Vols. 8–9, p. 296.

[9] Michael D. Lemonick (2008). *The Georgian star: how William and Caroline Herschel revolutionized our understanding of the cosmos*. W.W. Norton & Company.

January 9, 1848. In the obituary published by the Royal Astronomical Society we can read[10]:

> . . . sister to the celebrated astronomer of that name, as well as the con-
> stant companion and sole assistant of his astronomical labours, to the
> success of which her indefatigable zeal, diligence, and singular accuracy
> of calculation, not a little contributed. [. . .]
>
> At her funeral, which took place on the 18th of January, the coffin was
> adorned with palm branches by order of the Princess Royal, and followed
> by a royal carriage. Her memory will live, with that of her brother, as
> long as astronomical records of the last and present century are pre-
> served; and it will live on its own merits, even though, as may reasonably
> be hoped, the time should come when the astronomical celebrity of a
> woman will not, by the mere circumstance of sex, be sufficient to excite
> the slightest remark. It must be matter of congratulation that she sur-
> vived to see the enormous undertaking commenced by her brother
> extended and perfected by her nephew.

Her crater is small, 13.4 km in diameter, in the western part of Mare Imbrium, south of Sinus Iridum. It was dedicated to her in 1824 by the selenologist Wilhelm Lohrmann.

---

[10] Report of the Council to the Twenty-eighth Annual General Meeting (1848). *Monthly Notices of the Royal Astronomical Society*, Vol. 8, Issue 4, February 11, p. 64.

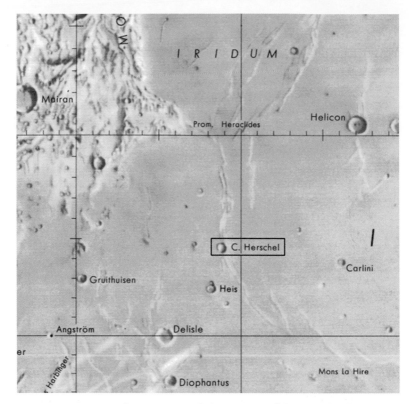

Figure 26 Location of crater C. Hershel. Courtesy of the Lunar and Planetary Institute, Houston, Texas.

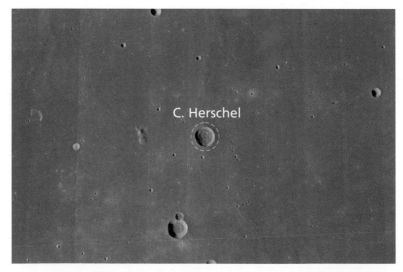

Figure 27 Lunar Reconnaissance Orbiter zoom on crater C. Hershel (image width is 150 miles).

5

Mary Fairfax Greig Somerville (1780–1872)

Figure 28 Mary Somerville (Thomas Phillips 1834, Scottish National Portrait Gallery).

> To discern and deduce from ordinary and apparently trivial occurrences the universal laws of nature as Galileo and Newton have done is a mark of the highest intellectual power.
>
> MARY SOMERVILLE[1]

In the May 5, 1860 issue of the literary and cultural journal *Atlantic Monthly* (founded in 1857 and published in Boston), we read: "There have been in every age a few women of genius who have become the successful rivals of man in the paths which they have severally chosen. Three instances are of our time. Mrs. Browning is called a poet even by poets; the artists admit that Rosa Bonheur is a painter, and the mathematicians accord to Mary Somerville a high rank among themselves."

Mary Somerville was known in her time as the "Queen of Science"— not just a mathematician, not just an astronomer, she was described by Dorothy McMillan as, "Feminist, Mathematician, Astronomer, Botanist, Geographer – Scientist." This was in McMillan's introduction to Somerville's *Personal Memoirs*[2], published shortly after her death by her

---

[1] Kathryn A. Neeley (2001). *Mary Somerville: science, illumination, and the female mind.* Cambridge University Press.

[2] Dorothy McMillan (ed.) (1873, 2009). *Mary Somerville, queen of science: personal recollections of Mary Somerville.* Canongate Classics.

The Women of the Moon. Daniel R. Altschuler Stern and Fernando J. Ballesteros Roselló.
© Daniel R. Altschuler Stern and Fernando J. Ballesteros Roselló 2019. Published in 2019 by Oxford University Press. DOI: 10.1093/oso/9780198844419.001.0001

daughter Martha in 1873. In light of this praise, it is somewhat surprising to read Somerville's own self-evaluation, the opening of those memoirs: "My life has been domestic and quiet. I have no events to record that could interest the public, my only motive in writing it is to show my countrywomen that self-education is possible under the most unfavorable and even discouraging circumstances."

Accustomed, as we are, to seeing history—in particular, the history of science—as a sequence of important milestones, new theoretical, observational, or experimental discoveries associated with names and surnames, it becomes difficult to catalog Somerville and to understand her significance. She did not discover anything new, she did not develop new theoretical ideas, not even, as was the case of her contemporary Caroline Herschel (1750–1847), was she a key contributor to developments credited to others.

She confesses, in the manuscript of her long autobiographical essay (but not published in the final version)[3]: "In the climax of my great success, the approbation of some of the first scientific men of the age and of the public in general I was highly gratified, but much less elated than might be expected, for although I had recorded in a clear point of view some of the most refined and difficult analytical processes and astronomical discoveries, I was conscious that I had never made a discovery myself, that I had no originality."

Despite her own evaluation, Somerville was an important and singular figure in the history of science and in her time, a prominent public personality admired and appreciated by her contemporaries.

She was born on December 26, 1780, at the home of her uncle, the Reverend Dr. Thomas Somerville (1741–1830), in the town of Jedburgh, in the Scottish borderlands, and was raised in the port town of Burntisland, on the shore of the Firth of Forth. She was one of the seven children of Sir William George Fairfax (1739–1813), Vice Admiral of the Royal Navy, and Margaret Charters Fairfax (1741–1832).

As was customary, her education was limited to what was considered proper for a girl. Its aim was to produce a good wife, a good home-keeper, and a good mother. Curious, by fourteen she had read the few books available in her house: Shakespeare (against the objection of the other women in the family) and some books on navigation belonging

[3] Kathryn A. Neeley (2001). *Mary Somerville: science, illumination, and the female mind*, p. 188. Cambridge University Press.

to her father. By her teen years, her family was spending winters in the capital, Edinburgh, where Mary could go to a proper school and learn writing and arithmetic, drawing, and painting. At thirteen, her uncle, the Reverend Somerville, taught her Latin.

Her curiosity was truly awakened by some mathematical riddles she saw in a magazine. She asked her brother's tutor to get her John Bonnycastle's algebra book and Euclid's geometry book, the standard texts used in schools (at that time, it was unthinkable for a woman to enter a bookstore). Upon discovering this, her father forbade her to read them (but apparently, he did not confiscate them), fearing that these readings would affect her femininity and the mental effort could drive her crazy, a widespread opinion in those days. But as so often happens, the forbidden became even more attractive, and young Mary studied her books secretly.

In 1804, at twenty-four, she married a second cousin, Captain Samuel Greig, and they moved to London, where Samuel served as emissary and commissioner of the Russian navy. During the first years of her marriage, Mary studied French and mathematics and bore two children. Samuel, like Mary's father, did not approve of her intellectual pretensions but she continued nevertheless. Luckily for Mary, Samuel died in 1807.

Her widowhood and an inheritance gave her the freedom she needed to continue her studies. She read Newton's *Philosophiæ Naturalis Principia Mathematica*, one of the seminal mathematical texts since its publication in 1687, and she studied astronomy. Her family and her friends objected, but she found support in the Scottish mathematician and geologist John Playfair (1748–1819) and his disciple William Wallace (1768–1843), future professor of mathematics at the University of Edinburgh and editor of a mathematical magazine for general readers. At the suggestion of the latter, she acquired a collection of mathematical and scientific books, among them Pierre-Simon de Laplace's *Mécanique Celeste,* and began to study. For Somerville, this small library was a treasure.

She was also encouraged by her cousin, the surgeon William Somerville (1771–1860), son of the Reverend Somerville, who after a long absence abroad—as far as the exploration of southern Africa—returned to Scotland in 1811. William and Mary were married in 1812 and went to live in Edinburgh, where he became the director of army hospitals. In a sharp contrast with her first husband (and with most husbands of the time), William supported Mary in her studies, and when in 1815 he was transferred to London they began to participate in the activities of the Royal Society and to associate with the most eminent scientists of the time, who received the not only brilliant, but affable and beautiful Mary with respect and friendship. In 1817, the Somervilles visited France, Switzerland, and Italy, where they were well received by local scientists. In Paris, she met, among others, the zoologist and naturalist Georges Couvier (1769–1832), the physicist and mathematician Siméon Poisson, the physicist Felix Savart (1791–1841), the geographer and explorer Alexander von Humboldt (1769–1859), and none other than the eminent Pierre-Simon Laplace (1749–1827), who had learned that she had read his treatise. Laplace is said to have commented that, in addition to Mary Somerville, only two other women had understood his work: a Miss Fairfax and a Mrs. Greig, of whom he knew nothing[4].

After this long and pleasant journey, the Somervilles returned to England in 1818.

In 1826, she published the results of an experiment that studied the magnetization of a needle exposed to sunlight. Although the results were later overturned (light does not magnetize), the work was well received by the community of natural philosophers and was the first article published by a woman in the *Philosophical Transactions of the Royal Society*. The astronomer John Herschel (several times president of the Royal Astronomical Society, son of William Herschel and nephew of Caroline), the mathematicians Charles Babbage (1792–1858) and George Peacock (1791–1858), and the scientist William Whewell[5] (1794–1866) were part of her circle of friends. Ultimately, however, Mary did not distinguish herself so much as a scientist, but rather as a popularizer of science.

In 1827, she was asked by Lord Henry Brougham to write a summary of Laplace's *Celestial Mechanics* to be published by the Society for the Diffusion of Useful Knowledge, over which he presided. Mary accepted with two conditions: that the project be kept secret and that, if it was

---

[4] In all likelihood, a witty legend. Its apocryphal nature is echoed in a story about the distinguished astrophysicist Sir Arthur Eddington (1882–1944) who was similarly told: "Professor Eddington, you are one of only three people in the world who understand Einstein's general theory of relativity." Upon Eddington's silence, it was added, "Say, Eddington, do not be modest," to which he replied: "I was trying to think who the third person is."

[5] Whewell, incidentally, coined the term "scientist" in 1833 to replace the former "natural philosopher," and also suggested to Faraday the terms "anode" and "cathode."

unsatisfactory, the manuscript would be burned. That commission was a crucial step for her career.

After four years of work, and with a recommendation by John Herschel, *The Mechanisms of the Heavens* was published in 1831 by the publisher John Murray, as the work far exceeded the space available for the publication by the Society for the Diffusion of Knowledge. Both Murray and Somerville were surprised by the book's positive reception, with enthusiastic reviews on both sides of the English Channel. This work was much more than a translation or summary. Mary added explanatory material, updated it with new results, put Laplace's work in a historical context, and wove a narrative consonant with a vision of natural theology—that the study of nature allows us to understand its divine design. The reader is told that the effort will allow him to contemplate the most sublime works of the Creator[6].

Contrast this with Laplace's famous response to Napoleon's question about where the Creator was in his mechanics: "I do not need that hypothesis." For a century, Somerville's book persisted as a text in university courses in mathematics and astronomy.

Something similar had happened almost a hundred years earlier, when Émilie du Châtelet (1706–1749, friend and lover of Voltaire) undertook the task of translating Newton's *Philosophiae Naturalis Principia Mathematica* from Latin and preparing a commented French version. She finished this work shortly before dying from complications in childbirth in 1749. It is the only complete translation of Newton's work available in French and still considered the reference version in this language (Émilie, alas, does not have a crater).

In 1834, Somerville published her second book, *On the Connection of the Physical Sciences*, which was even better received than the first and well-nigh consecrated its author. The eminent physicist James Clerk Maxwell, creator of the field of electrodynamics, commented that "it is a suggestive book, that puts in a definitive, intelligible, and communicable way, the ideas that are already inhabiting the minds of men [sic] of science, guiding them to discoveries that they cannot yet formulate clearly." In each successive edition of the book, Somerville updated it with the most recent discoveries. In the fifth edition of 1840, for example, we can read about the problem presented by the planet Uranus, which did not conform to its expected orbit: "The tables of Uranus, however, are already

---

[6] Mary Somerville (1831). *Mechanism of the heavens*, p. 144. Murray.

defective probably because the discovery of this planet in 1781 is too recent as to admit of great precision in the determinations of its movement or because possibly it is subject to the disturbance by some invisible planet in orbit around the Sun, beyond the present limits of the system."

As if to illustrate Maxwell's point, it is likely that this note on Uranus's orbit was read by John Couch Adams (1819–1892), then a student of mathematics at Cambridge. He began work on the problem after 1843, finding a solution in 1845. He gave the results of his calculations, indicating the location of the potential new planet, to two important English astronomers: James Challis (1803–1882), director of the Cambridge Observatory, and George Airy (1801–1892), royal astronomer and director of the Royal Greenwich Observatory. Neither of them took the initiative to point a telescope at the region indicated by Adams. Meanwhile, Uranus was also the subject of study in Paris, where Urbain Le Verrier began working on the problem in 1845, and by mid 1846 had obtained a position not unlike that obtained by Adams. This work was published and also sent to Airy. Finally, nine months after receiving Adams's data about where to look for the planet, and convinced by the results of Le Verrier, Airy asked Challis to start the search with a Cambridge telescope.

During July and August of 1845, the positions of all the stars in the area were recorded, searching for one whose position changed, indicating a possible planet. Meanwhile, Adams and Le Verrier, unaware of each other's work, refined their calculations and obtained better predictions of the position of the new planet, which did not differ much from each other (Adams calculated a longitude in the ecliptic of 330° and Le Verrier of 326.5°). Looking for some telescope available for research, Le Verrier wrote to Gottfried Galle (1812–1910), at the Berlin Observatory, who obtained permission from the director, Johann Encke (1791–1865), to search. Thus, on September 23, 1846, Galle, assisted by his student Heinrich d'Arrest (1822–1875), comparing the visible stars with maps of the sky, discovered one that was not on the maps, at a longitude of 327°: Neptune[7]. A reanalysis of Challis's observations, as it turns out, showed that he had seen Neptune in August, and today he is remembered for having missed this great opportunity.

---

[7] It was a matter of luck that the course of their decades-long orbits brought Uranus and Neptune close enough to each other so as to disturb each other, causing the irregularities that inspired the search.

Let's get back to Mary Somerville.

In 1835, Mary and Caroline Herschel were elected as honorary members of the Royal Astronomical Society, the first women to receive this honor. By then, Caroline had already returned to her hometown of Hanover, and she received with great happiness a letter from Somerville in which she confided that the honor was even more valuable for being shared with Herschel, whom she greatly admired. In 1838, William Somerville became ill, and decided to retire to the warmer climate of Italy; the Somervilles moved to Florence with their daughters, Martha and Mary. They lived in several Italian cities for the rest of their lives. In 1848, working in Florence, Mary published *Physical Geography* and in 1869 *Molecular and Microscopic Science*. Until the end of his life, William encouraged his wife. In her memoirs, Mary tells us: "The warmth with which Somerville shared my success affected me deeply, since not one in ten thousand would have celebrated as he did, he was generous in nature, far above jealousy, and continued all his life to be generously interested in everything I did." William Somerville died in Florence in 1860.

Kathryn Neeley, in her biography of Somerville, which is also an excellent text on the subject of women and science, writes:

> At a time when science was perceived as a man's domain, Mary Somerville was the most important scientific woman of the time and an integral part of the British scientific community. Her scientific treatises were important cultural contributions to Victorian England and served to establish science as an integral and unifying distinctive element of culture. At the time of her death, Somerville occupied an almost mythical position in Britain. Her work reflects the power of science to capture the imagination and the influence of cultural factors in the development of science. They provide a window to a particularly lucid and enlightened mind and to one of the most formative periods in the evolution of modern scientific culture.

Mary Somerville, for her own part, put it as follows[8]:

> Age has not abated my zeal for the emancipation of my sex from the unreasonable prejudice too prevalent in Great Britain against a literary and scientific education for women. The French are more civilized in this respect, for they have taken the lead, and have given the first example in modern times of encouragement to the high intellectual culture of the

---

[8] Dorothy McMillan (ed.) (1873, 2009). *Mary Somerville, queen of science: personal recollections of Mary Somerville*, p. 278. Canongate Classics.

sex. Madame Emma Chenu, who had received the degree of Master of Arts from the Academy of Sciences in Paris, has more recently received the diploma of Licentiate in mathematical science form the same illustrious society . . . A Russian lady has also taken a degree, and a lady of my acquaintance has received a gold medal from the same institution. (She refers to Sophie Germain.)

At eighty-nine, she contemplated the approach of death:

Yet to me, who am afraid to sleep alone on a stormy night, or even to sleep comfortably any night unless someone is near, it is a fearful thought, that my spirit must enter that new state of existence quite alone. We are told of the infinite glories of the state, and I believe them, though it is incomprehensible to us; but as I do comprehend, in some degree at least, the exquisite loveliness of the visible world, I confess I shall be sorry to leave it. I shall regret the sky, the sea with all the changes of their beautiful colouring; the earth, with its verdure and flowers; but far more shall I grieve to leave animals who have followed our steps affectionately for years, without knowing for certainty their ultimate fate.

By the age of ninety-two, she seemed to have achieved peace with the idea of mortality:

The Blue Peter [the maritime banner that indicates that a ship is putting to sea] has been long flying at my foremast, and now I am in my ninety-second year I must soon expect the signal for sailing. It is a solemn voyage, but it does not disturb my tranquility. Deeply sensible of my utter unworthiness, and profoundly grateful for the innumerable blessings I have received, I trust in the infinite mercy of my Almighty Creator. I have every reason to be thankful that my intellect is still unimpaired, and, although my strength is weakening my daughters support my tottering steps, and, by incessant care and help, make the infirmities of age so light to me that I am perfectly happy.

And so it was. Mary Somerville died in 1872 at ninety-two. Her tomb, together with that of her two daughters, Mary who died in 1875 and Martha who died in 1879, is located in the *Cimitero degli Inglesi* in Naples. In 1865, William R. Birt and John Lee had baptized a crater in her honor, but in 1976, the IAU named a different one. This small crater of 15 km is on the visible side to the right of Mare Fecunditatis (Sea of Fertility), next to the large crater Langrenus and very close to the edge of the visible disk.

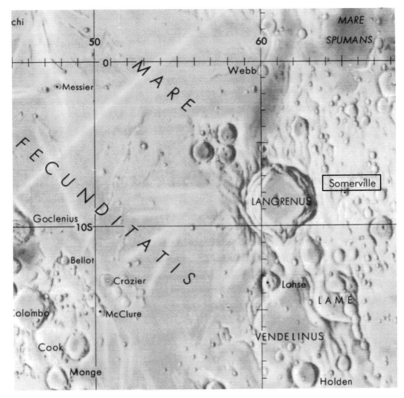

Figure 29 Location of crater Somerville. Courtesy of the Lunar and Planetary Institute, Houston, Texas.

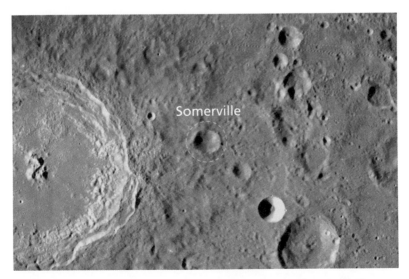

Figure 30 Lunar Reconnaissance Orbiter zoom on crater Somerville (image width is 150 miles).

6

Anne Sheepshanks (1789–1876)

In an 1876 obituary written by the Royal Astronomical Society of England, we read[1]:

> Our late Honorary Member, Miss Anne Sheepshanks, who died at an advanced age a few days before the last anniversary meeting, was one of the many patrons of our science whose memory will remain associated with the history of astronomy in the nineteenth century. The sister and companion of the late Rev. Richard Sheepshanks, of whom she was the senior by five years, she shared the interest which he always took in the progress of astronomy, and especially in the proceedings of this Society, in the administration of which he had labored so faithfully and successfully to within a short period of his death.

We know relatively little about the life of Anne Sheepshanks; most of what we do know is in the context of her brother, Richard Sheepshanks (1794–1855), a relatively minor astronomer[2] who from 1829 until his death was secretary of the Royal Astronomical Society and editor of its journal: *Monthly Notices*. Born to a wealthy family, they were the children of Joseph Sheepshanks, a textile manufacturer in Leeds (Yorkshire), and his wife Anne. The siblings lived together (her obituary describes her as the sister and companion of Reverend Richard), and neither of them married.

Anne was the only heir to her brother, (who, despite being a reverend, had six children with an Irish dancer, to whom he seems to have left not a penny), and in 1858 she donated the considerable sum of 10,000 pounds (equivalent in purchasing power to about a million dollars today) to the University of Cambridge to promote research in astronomy, terrestrial magnetism, and meteorology. In 1860, she donated another

---

[1] Edwin Dunkin (1879). *Obituary notices of astronomers, fellows and associates of the Royal astronomical society written chiefly for the annual reports of the council*. Williams and Norgate.

[2] One notes that no crater is named for Richard Sheepshanks. Money, in this case, must have spoken louder than his service to the Royal Astronomical Society.

The Women of the Moon. Daniel R. Altschuler Stern and Fernando J. Ballesteros Roselló.
© Daniel R. Altschuler Stern and Fernando J. Ballesteros Roselló 2019. Published in 2019 by Oxford University Press. DOI: 10.1093/oso/9780198844419.001.0001

2000 pounds for the acquisition of a twelve-and-a-half-inch refractor telescope for the observatory, though this was not built until 1898. The
Sheepshanks telescope will stand in for her portrait, as there are no
known extant images of her person. (The telescope, it must be admitted,
turned out to be a rather unsatisfactory instrument. As described by
Cecilia Payne-Gaposchkin[3], in the words of William Marshall Smart, it
"combined all the disadvantages of a refractor and a reflector.") In 1857,
Anne donated Richard's valuable collection of instruments and books to
the Royal Astronomical Society, and in 1862 was named an honorary
member of the society (honors granted only to Caroline Herschel and
her contemporary Mary Somerville).

Her obituary states that after the death of her brother in 1855, Anne
lived a very retired life in her residence in Reading where she died in
1876 at the age of eighty-six years. One might note that there is no crater

Figure 31 The Sheepshanks telescope. Institute of Astronomy Library,
University of Cambridge.

[3] Cecilia Payne-Gaposchkin (1996). *An autobiography and other recollections*, p. 119. Cambridge
University Press.

in honor of Richard and it is not unreasonable to say that she "bought" her crater. In 1865, Birt and Lee, commissioned by the British Association, prepared a high-resolution map of the Moon based on Madler's map. They proposed an extensive list of names for the new craters they mapped. Of their proposed new names, only 85 are official today, one of them being Sheepshanks. Her crater, measuring 25 km in diameter, is at the northern end of Mare Frigoris (sea of cold), on the Moon's visible side, near the moon's north pole.

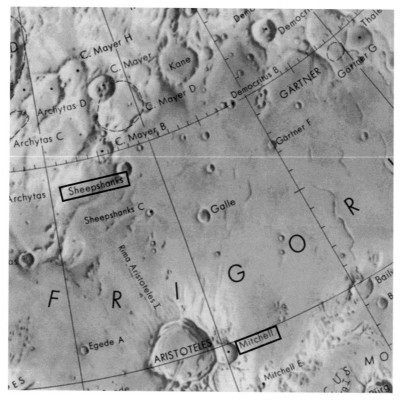

Figure 32 Location of craters Sheepshanks and Mitchell. Courtesy of the Lunar and Planetary Institute, Houston, Texas.

Figure 33 Lunar Reconnaissance Orbiter zoom on crater Sheepshanks (image width is 150 miles).

7

Catherine Wolfe Bruce
(1816–1900)

Figure 34 Possible portrait of Catherine Wolfe.

Such a blow from a friend! I think we are beginning—else why set to work [on] Photography, Spectroscopy, Chemistry and soon but perhaps not in this generation Electricity. Think of the great mechanical improvements, think of the double stars revolving around a common center, in variable stars. The world is young.

From a letter written on November 6, 1890, to Simon Newcomb, the renowned astronomer, and professor of mathematics at the Naval Observatory, in response to his 1888 article in which he argued that all important astronomical discoveries had already been made. (Newcomb Papers)

The life of Catherine Wolfe Bruce has certain parallels with that of Anne Sheepshanks, who turned twenty-six when Catherine was born in New York in 1816, the second daughter of George Bruce (1781–1866) and his wife Catherine Wolfe (1785–1861). George was a wealthy printer who made his fortune by being first to offer stereotype printing in America; he invested his earnings in real estate in New York City, and his family was one of the city's wealthiest of the nineteenth century. On his death, George left a fortune to his daughters.

Catherine, although influential and recognized in certain circles, lived a very reserved life, a wealthy, highly educated spinster of delicate health, accompanied by her younger sister Matilda, who administered her affairs. With a fortune in her pocket and no need to earn more,

The Women of the Moon. Daniel R. Altschuler Stern and Fernando J. Ballesteros Roselló.
© Daniel R. Altschuler Stern and Fernando J. Ballesteros Roselló 2019. Published in 2019 by Oxford University Press. DOI: 10.1093/oso/9780198844419.001.0001

Catherine devoted herself to traveling and enjoying art and literature. In 1877, she donated $50,000 (equivalent to about $1,200,000 today) to establish a branch of the New York Public Library in memory of her father. The George Bruce Branch Library is located on the south side of West 125th Street and lies within the boundaries of the General Grant Houses, erected during the Robert Moses' Title I urban renewal era (1949–1957), when it was common practice to locate public housing projects near other public entities.

We have no independent source of her portrait (which we found here: http://www.thepandorasociety.com/this-day-in-history-december-22nd-1891/), but Figure 35 shows a photograph of the Bruce telescope.

24 inch Bruce Photographic Telescope (at Cambridge)

Observatory

Figure 35 Twenty-four-inch Bruce Photographic Telescope at Cambridge. Harvard University Archives, UAV 630.271 (164).

How did there come to be a telescope (and ultimately, a crater) named after her? At the age of seventy-three, challenged by Newcomb's article, as our chapter epigraph shows, she took up supporting many astronomers, both in the United States and Europe, buying instruments as well as paying salaries and the costs of some publications. In a way, she was encouraging them to make Newcomb swallow his words with their deeds. When she came across an open letter in which Edward Pickering (1846–1919), director of the Harvard Observatory, requested $50,000 for a telescope, Bruce gave him the money and the twenty-four-inch telescope was completed in 1893. In 1896 the telescope was moved to the station that the Harvard Observatory had acquired in Arequipa, Peru, to observe the Southern Hemisphere's sky.

She continued to sponsor astronomy projects with over fifty donations between 1889 and 1899 totaling $275,000 (equivalent to about $7,000,000 today). In her list of donations, we see that she gave $100 to David Gill in Cape Town in South Africa (a friend of Agnes Mary Clerke, whom we will soon meet), $350 to Mary Whitney of Vassar, and $10,000 to Max Wolf of Heidelberg for the purchase of a telescope. With this instrument, Wolf discovered a new asteroid and named it "Brucia" in her honor.

Alongside Mrs. Draper of New York (wife of Henry Draper) and Mrs. Alice Bache Gould from Boston, who in 1897 donated $20,000 to the National Academy of Sciences to support astronomical research, Catherine was a leading benefactor of astronomy at the turn of the twentieth century. (Alas, there are no craters honoring them, although there is a crater for Henry Draper.)

If she is remembered today, it is for the medal that bears her name. In 1897, she donated $2750 to the Astronomical Society of the Pacific, to establish the Bruce Medal (or Catherine Wolfe Bruce Gold Medal, to use its official name), one of the most prestigious astronomical recognitions in the world. This gold medal is awarded annually by the Astronomical Society of the Pacific to astronomers from all over the world, recognizing the trajectory of their life's work (which is why it cannot be granted twice to the same person). Candidates are proposed by the directors of three American astronomical observatories and three others from the rest of the world. Of the initial $2750, $250 were destined to create the molds of the medal.

We note that Simon Newcomb (1835–1909) was the first winner of the Bruce medal in 1898. It was not until 1982 that it was awarded to a woman: E. Margaret Burbidge (born in 1919). To date, there have been only three other women among the 112 winners: Charlotte Emma Moore Sitterly (1898–1990), Vera Florence Cooper Rubin (1928–2016), and Sandra Moore Faber (born 1944).

Catherine Bruce died at her residence in New York. She was eighty-four years old. In a note published in 1927, H.B. Kastner writes[1]: "Before her death in New York on May 13, 1900, Miss Bruce was responsible for greater advances in astronomy than perhaps any other woman before or since. It was due perhaps to her intimate association with the greatest astronomers of the time, both in Europe and in America, that her many financial gifts were placed where they would result in the greatest benefit for science."

Two years before her death, the same year that the first Bruce medal was awarded, the selenographer Johann N. Krieger baptized a Moon crater with her name. Her small crater, 7 km in diameter, is very close to the one of Mary Adela Blagg, near the lunar point closest to Earth, from where it is always seen at the lunar zenith.

---

[1] Kaster H.B. (1927). Note regarding the death of Miss Catherine Wolfe Bruce and her services to astronomy. *Publications of the Astronomical Society of the Pacific*, Vol. 39, No. 231, p. 325.

Figure 36 Location of craters Bruce and Blagg. Courtesy of the Lunar and Planetary Institute, Houston, Texas.

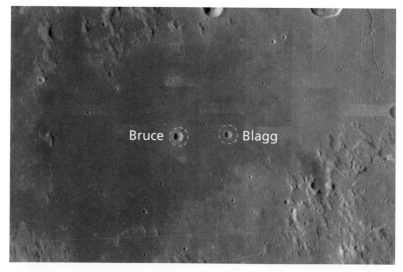

Figure 37 Lunar Reconnaissance Orbiter zoom on craters Bruce and Blagg (image width is 150 miles).

8

Maria Mitchell (1818–1889)

Figure 38 Maria Mitchell.
Archives and special collections
Vassar College.

The laws of nature are not discovered by accidents; theories do not come by chance, even to the greatest minds; they are not born of the hurry and worry of daily toil; they are diligently sought, they are patiently waited for, they are received with cautious reserve, they are accepted with reverence and awe. And until able women have given their lives to investigation, it is idle to discuss the question of their capacity for original work.

MARIA MITCHELL[1]

On October 1, 1847, a twenty-nine-year-old woman, a librarian on Nantucket island off the coast of Massachusetts (and a major whaling port), climbed onto the roof of her house, as was her habit, and with a small two-inch refractor telescope started to look at the sky. She noticed a blurry object that she had not seen before in that part of the sky, and thinking that it might be a comet, she went down to look for her father, William, another amateur astronomer. After checking the finding, he immediately reported it to a friend, William C. Bond (1789–1859), first director of the Harvard College Observatory in Boston, who confirmed the discovery. Thus, sixty-one years after Caroline Herschel discovered her first comet, Maria Mitchell became the first American woman to do so. It happened that the king of Denmark, Frederick VI (1768–1839), who was an amateur astronomer, had instituted a gold medal to recognize discoverers of new comets (in the rising age of democracy, kings still had

---

[1] Maria Mitchell (2001). *A life in journals and letters* (Henry Albers, ed.). College Avenue Press.

The Women of the Moon. Daniel R. Altschuler Stern and Fernando J. Ballesteros Roselló.
© Daniel R. Altschuler Stern and Fernando J. Ballesteros Roselló 2019. Published in 2019
by Oxford University Press. DOI: 10.1093/oso/9780198844419.001.0001

to do something), and after some deliberations about how the discovery had been communicated, the medal was granted to Maria. These events, widely reviewed in the international press, boosted the fame of the shy Maria. In 1848, she became the first woman to be elected to the American Academy of Arts and Sciences.

Let's go back to the year 1818, when Lydia Coleman Mitchell (1764–1840), wife of William Mitchell (1791–1869, school teacher), gave birth to her third child (seven more would follow), Maria. As Quakers (members of Religious Society of Friends), they subscribed to the idea that men and women deserved equal education, and Maria received the basic elements of primary education: reading, writing, and arithmetic. To know the sky was part of the local educational tradition: of great importance for navigators, it was natural for the inhabitants of sea-dependent Nantucket. With her father she studied astronomy and at an early age she assisted him with his job of calibrating chronometers for navigation.

In 1836, at the age of eighteen, she began working as a librarian at the Atheneum in Nantucket and took advantage of her free time to continue studying. She was a voracious reader consuming, among many other books that fell into her hands, Laplace's *Celestial Mechanics* and the *Theorie Motus Corporeum Coelestium* by Gauss—in Latin. Eminent men of science occasionally visited her father, and she too made their acquaintance, not merely socially interacting with these figures, she assisted her father with measurements of the position of thousands of stars for the United States Coast Survey, whose superintendent was William's friend. After her comet in 1847, she was hired by the United States Nautical Almanac to observe and catalog the position of Venus (useful for navigation), work that she did after her day job as a librarian. She was, in short, fascinated by the night sky and in her diary for 1854 we can read[2]: "Last night I examined the sky for two hours. It was a great night – not a breath of air, not a fringe of a cloud, all clear, all beautiful. I really enjoy that kind of work, but my back soon becomes tired, long before the cold chills me. I saw two nebulae in Leo with which I was not familiar and that repaid me for the time."

In 1857, she accompanied a friend, a banker's daughter, on a trip to Europe. At Westminster Abbey in London she visited Newton's tomb, and she met the royal astronomer Sir George Airy and the prominent

---

[2] Maria Mitchell (2012). *Life, letters and journals*, p. 29. Forgotten Books.

Sir John Herschel, Caroline's nephew, who wrote her a letter of introduction to visit Mary Somerville. The banker's daughter returned to
the United States, but Maria continued her trip. In Rome, she got permission to visit the Vatican Observatory (no women were allowed and
neither Somerville nor the daughter of John Herschel had managed to
visit). In her diary she wrote[3]:

> I said to myself, 'This is the land of Galileo, and this is the city in which he
> was tried. I knew of no sadder picture in the history of science than that
> of the old man, Galileo, worn by a long life of scientific research, weak
> and feeble, trembling before that tribunal whose frown was torture, and
> declaring that to be false which he knew to be true. And I know of no
> picture in the history of religion more weakly pitiable than that of the
> Holy Church trembling before Galileo, and denouncing him because he
> found in the Book of Nature truths not stated in their own Book of God
> – forgetting that the Book of Nature is also a Book of God.

In January 1858, she traveled to Florence, where she visited the elderly
Mary Somerville. Of her, Mitchell wrote admiringly in a letter to her
father[4]:

> I could but admire Mrs. Somerville as a woman. The ascent of the steep
> and rugged path of science had not unfitted her for the drawing-room
> circle; the hours of devotion to close study have not been incompatible
> with the duties of wife and mother; the mind that has turned to rigid
> demonstration has not thereby lost its faith in those truths which figures
> will not prove. "I have no doubt," said she, in speaking of the heavenly
> bodies, "that in another state of existence we shall know more about
> these things."

After Italy she went to Berlin to visit the illustrious Alexander von
Humboldt. She returned after two years overseas to Nantucket, without imagining that a wealthy brewer would change her life.

Matthew Vassar (1792–1868), born in England, emigrated as a child
with his family to the United States in 1796, and prospered as owner of
M. Vassar & Co., a brewery. In 1861, Vassar presented $408,000 in a tin
box to the trustees of a new college[5] he wanted established and to whom
he declared: "It occurred to me, that woman, having received from her

---

[3] Maria Mitchell (2012). *Life, letters and journals*, p. 151. Forgotten Books.
[4] Maria Mitchell (2012). *Life, letters and journals*, p. 163. Forgotten Books.
[5] http://vcencyclopedia.vassar.edu/matthew-vassar/more-than-a-brewer.html

Creator the same intellectual constitution as man, has the same right as man to intellectual culture and development." Construction began on the one of the first institutions of higher education for women in the Western world. Located in Poughkeepsie, New York, Vassar Female College opened its doors in 1865 (the word "female" was removed from its name in 1867).

On June 23, 1868, Vassar, at the age of seventy-four, presented his farewell message (never better said) to the board of trustees, as it was his intention to resign. The minutes of the meeting document that fateful moment: "[he]

Figure 39 Matthew Vassar presents a gift for Vassar College.

was reading from the eleventh page when he failed to pronounce a word which was upon his lips, dropped the papers from his hand, fell back in his chair insensible, and died at precisely ten minutes to 12 o'clock AM by the clock in the College Tower." After a respectful interval, and a prayer, the board reconvened to hear the chairman read the rest of Vassar's remarks[6] which towards the end state[7]: "And now, gentlemen, in closing these remarks, I would humbly and solemnly implore the Divine Goodness to continue his smiles and favor on your institution, and bestow upon all hearts connected therewith his love and blessings, having peculiarly protected us by his providence through all our college trials for three consecutive years, without a single death in our Board or serious illness or death of one of our pupils within its walls." *The New York Times* reported on June 24, 1868: "It is a singular circumstance, that at the time of his death, he was attired in an entirely new suit, funeral black, which was not removed, and he will be buried in it in accordance with the wishes of his family."

---

[6] Allison Guertin Marchese (2017). *Hudson Valley curiosities*, p. 82. The History Press.
[7] http://vcencyclopedia.vassar.edu/early-vassar/communications-to-the-board-of-trustees/11-june-23-1868.html

Among the nine teachers (two of them women) who initially formed Vassar's faculty was Maria Mitchell, director of the new Vassar College Observatory, with a respectable twelve-inch telescope, at that time the third largest in the United States[8]. Soon, the "Mitchell girls" made measurements from the observatory and on organized trips to contemplate solar eclipses. Mitchell herself began pioneering work on sunspot photography, and published work on the changes observed on the surfaces of Jupiter and Saturn.

Three of her students were named in the 1906 first edition of the *American Men of Science* directory (despite not being "men"): Antonia Maury (who we will meet soon); Mary Whitney (1847–1921), who went on to direct the observatory after the retirement of Maria; and the logician Christine Ladd-Franklin (1847–1930).

To briefly digress: Ladd was the first woman to obtain a PhD from Johns Hopkins. She signed her application for a scholarship as C. Ladd. When, after approval, the university authorities realized that Ladd was a woman, they tried to rescind it, but in the end, they allowed her to study and complete all the requirements for the doctorate. However, since Johns Hopkins did not provide doctorates for women, she left the university without this title, which she did not receive officially until 1926, at seventy-eight years of age. Bertrand Russell (1872–1970) tells us the following amusing story[9]: "I once received a letter from an eminent logician, Mrs. Christine Ladd-Franklin, saying that she was a solipsist, and was surprised that there were no others. Coming from a logician and a solipsist, this surprise surprised me."

Let's return to Mitchell.

Mitchell established a tradition that continues to the present: the Festival of the Dome, an event in which she brought distinguished guests to the observatory to discuss extracurricular questions with students, such as the issue of women's rights, a red-hot topic at the time. It was also customary for each participant to compose verses.

In 1873, she became a founding member and president of the Association for the Advancement of Women, a group of feminists who met annually to discuss women's issues and promote equal rights, and met in this capacity with Susan B. Anthony.

---

[8] At present, the telescope with the largest mirror is the Gran Telescopio Canarias (GTC), whose diameter is 10.4 meters.

[9] Bertrand Russell (1948). *Human knowledge: its scope and value*, p. 161. Routledge.

As educator and activist, it is worth considering Mitchell's career and legacy in light of the century of history that has followed. Renée Bergland offers us such a summary in a book written ten years ago[10]: "Still, there is great hope in Mitchell's story. It is a happy fact that Maria Mitchell's career was possible, and it is heartening to know that there was a time in America, however brief, when girls were strongly encouraged to pursue sciences. We currently live in an era when the bias against women in science seems an eternal constant."

Shortly before her death she received a letter from a student that said: "In all the great wonder of life, you have given me more than I have desired, more than any other creature has given me. I was hoping to become something for you." We think that for an educator there is no medal that can improve on that real prize. Her crater, one of those named in 1865 by Birt and Lee, is located to the south of Sheepshanks and to one side of Aristotle, south of Mare Frigoris. You can find it on the map for Sheepshanks, shown in Figure 32.

Figure 40 Lunar Reconnaissance Orbiter zoom on crater Mitchell (image width is 150 miles).

[10] Renée Bergland (2008). *Maria Mitchell and the sexing of science: an astronomer among the American romantics.* Beacon Press.

9

Agnes Mary Clerke (1842–1907)

Figure 41 Agnes Mary Clerke (photo taken in 1905).

The life of a science is in the thought that binds together the facts; decadence has already set in when they come to be regarded as an end in themselves. "Man is the interpreter of nature"; to draw up an inventory, however, is not to interpret. It is true that speculation is prone to wander into devious ways: but then "truth emerges more easily from error than from confusion"

AGNES CLERKE, *The System of the Stars* (1890)

In her day, Agnes Clerke was well known as a popularizer of astronomy, a worthy successor to Mary Somerville[1]. She was born in the small town of Skibbereen on the south coast of Ireland, the daughter of John Willis Clerke, a bank manager (*c*.1814–1890), and his wife, Margaret (*c*.1819–1897).

John Willis, an educated man with a well-endowed library, taught his daughters Greek and Latin, as well as mathematics and science. In addition to his work at the bank, he dedicated himself to measuring time and calibrating clocks by observing the transit of stars (as did Maria Mitchell's father). This was of vital importance at a time when there were no electronic means of coordinating clocks over great distances (at that time astronomers were good for something practical!). Agnes was an avid reader; she would report in a later interview that she also had an immense love for astronomy, and love breeds knowledge.

[1] Mary Brueck (2002). *Agnes Mary Clerke and the rise of astrophysics.* Cambridge University Press.

The Women of the Moon. Daniel R. Altschuler Stern and Fernando J. Ballesteros Roselló.
© Daniel R. Altschuler Stern and Fernando J. Ballesteros Roselló 2019. Published in 2019 by Oxford University Press. DOI: 10.1093/oso/9780198844419.001.0001

At age fifteen, suffering from poor health, she and her sister went to live in Italy with their mother, in search of a more favorable climate. By then, Clerke had already developed a great interest in the history of astronomy, and with the help of her father she studied John Herschel's *Outlines of Astronomy*. In Italy, Agnes Mary studied science history and her sister, Ellen Mary (1840–1906), Italian literature. Ellen would become a distinguished journalist, essayist, and poet, publishing many articles in the *Dublin Review*, a Catholic magazine.

Returning to London in 1877, Agnes wrote two articles published in the *Edinburgh Review*, one of the most influential English publications in the nineteenth century. Starting her career as a writer on a broad range of topics, her first two articles were about the Italian Mafia and Copernicus in Italy, respectively. After writing an article on the chemistry of the stars, she decided to concentrate on astronomy, composing a total of fifty-five articles for the *Edinburgh Review*.

In 1885, after more than four years of work, she published her first book: *A Popular History of Astronomy during the Nineteenth Century*, five hundred pages long, very complete, and very well received internationally. Four editions were published, the last one in 1904. On its publication, many commentators already compared her to Mary Somerville. One wrote[2]:

> The book will be full of interest to the general reader, for the story of the marvelous discoveries made in astronomy during the past hundred years is told in a felicitous and attractive manner; but it will not be less highly valued by the student and the astronomer, on account of its completeness and accuracy, and the really remarkable skill with which the leading points on which our knowledge has been increased. . . are seized upon and set forth.

The author of these lines was Edward Maunder, husband of Annie Maunder, whom we shall soon meet.

The Scottish astronomer David Gill (1843–1914), Astronomer to His Majesty and Director of the Royal Observatory branch at the Cape of Good Hope, was a good friend of Agnes. The main mission of the Cape observatory was to complement the work of the Royal Observatory at Greenwich with observations from the Southern Hemisphere. In one of Gill's trips to organize an international project photographing the sky (*Carte du ciel*), he gave a lecture in London on the new subject of

[2] Mary Brueck (2002). *Agnes Mary Clerke and the rise of astrophysics*, p. 45. Cambridge University Press.

astrophotography. Following the presentation, Agnes wrote an article on "Sidereal photography." Thereafter, the friendship between Agnes, Gill, and his wife Bella strengthened. Noting that Agnes lacked practical experience in astronomy which would help with her work as an astronomical commentator, Gill invited her to spend a few months at the Cape Observatory:

> You are not complete until you have seen and done a little practical astronomy. Your work would take on a new and higher character after a little practical knowledge . . . For our sakes, for your own, and for the cause of astronomy I beg you to come. It Will do you the world of good and me a world of good also, just to have a real good talk about all the things you are in the midst of.

Agnes hesitated, since it was difficult for her to interrupt her writing work, and even harder to leave her family for so long, since, in her words, her elderly relatives were "fragile."

She was finally persuaded and embarked for South Africa on August 9, 1888, arriving on August 30. For two months, Agnes studied stars manifesting peculiar spectra and her results were published in the observatory journal. On October 31, she embarked for the voyage back to England.

In 1890, she published *The System of the Stars*, and in 1895, *The Herschels and Modern Astronomy*. In this book she emphasizes the contributions of Caroline beyond being the mere assistant to her famous brother.

In one of her books (*Modern Cosmogonies*) she writes[3]: "Year by year, details accumulate and the strain of keeping them under mental command becomes heavier." It is easy to forget that for her there were no resources such as those that make available to the modern writer all types of information easily accessible by electronic means. Orbital calculations are not the only part of astronomy to have benefited from modern computational technology!

In 1903, she published *Problems in Astrophysics*, a work so important that in that same year she was elected (along with Lady Huggins) as an honorary member of the Royal Astronomical Society, joining Caroline Herschel and Mary Somerville. Agnes's writing inspired many astronomers with whom she communicated frequently, often including their most novel results in her essays and books. She acted as if she were the

---

[3] Agnes Mary Clerke (1905). *Modern Cosmologies*, p. 160. Adam & Charles Black. https://archive.org/stream/moderncosmogonie00clerrich#page/160/mode/1up/search/year

director of an orchestra of astronomers since, on many occasions, she suggested new directions of investigation; many listened attentively. Gill told her about *Problems in Astrophysics*: "I do not believe there is a man living who knew beforehand all the facts you have brought together and brought together so well in their proper places."

Sometimes she gave free rein to her imagination. As a result of new results that indicated that the Sun was moving through our Galaxy, she wrote[4]:

> Each year, accordingly, we explore a belt of space nearly 400 million of miles in width, and our travelling, like that of the clouds, is "irrevocable." Shall we find ourselves in an "ampler ether" as we proceed? Or will the wreckage of our little planet help stock the void with meteorites? It may be that the gulfs will wash us down: It may be we shall touch the Happy Isles. Even the poets scarcely knew for certain which fate overtook Ulysses when he "sailed beyond the sunset" into a newer world.

The Sun, in its orbit around the galactic center, travels at a speed of about 800,000 km/hour and completes a revolution in about 225 million years. We can say that the age of the Earth is about 20 "galactic years."

Her beloved sister, Ellen, died in March 1906 at sixty-five years of age, after a flu that led to pneumonia. Ten months later, at the age of sixty-four, Agnes died of the same cause. Pneumonia, along with tuberculosis and gastrointestinal ailments, were the three leading causes of death in those times[5].

The tiny (7 km diameter) Clerke lunar crater, named by the IAU in 1973, is located near the landing site of Apollo 12, at the eastern end of the Mare Serenitatis, in the midst of a rille system named the Rimae Littrow.

---

[4] Mary Brueck (2002). *Agnes Mary Clerke and the rise of astrophysics*, p. 147. Cambridge University Press.

[5] The various types of pneumonia (bacterial and viral) presently afflict some four hundred and fifty million people each year but cause the death of only one percent of patients (most of them in non-developed countries), thanks to modern therapies and the use of antibiotics.

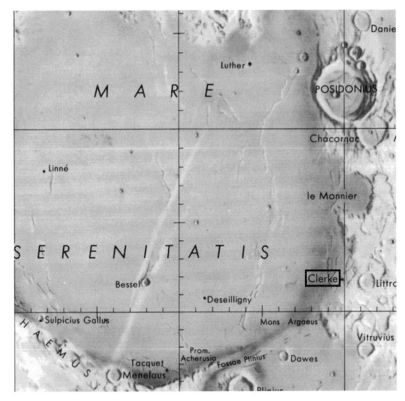

Figure 42 Location of crater Clerke. Courtesy of the Lunar and Planetary Institute, Houston, Texas.

Figure 43 Lunar Reconnaissance Orbiter zoom on crater Clerke (image width is 150 miles).

10

Sofia Vasílyevna Kovalévskaya (1850–1891)

Figure 44 Sofia Kovalévskaya (photo of 1880).

Oh, what a happy time it was! ... We were so enthusiastic about the new ideas; so sure, that the present social state could not continue long. We pictured to ourselves the glorious period of liberty and universal enlightenment of which we dreamed, and in which we firmly believed.

Sofia Kovalévskaya[1]

Decade after decade, century after century, the Moon always shows the same immutable face, while on this human timescale, the face of our society is ever-changing: slowly in the past, rapidly in the present. A hundred and fifty years ago, we lived in a world without computers or mobile phones, without cars or planes, without television or radio, without antibiotics or vaccines[2], without nuclear weapons or smart missiles, and so many other things that seem so normal to us. It was also a world without the threat of anthropogenic climate change. Hard to imagine, right?

But it was a world in which new revolutionary ideas arose; imaginations of the future and past were alive! Darwin published his fundamental *Origin of Species* in 1859, and Marx *Das Kapital* in 1867. In Eastern Europe, the winds of change were particularly strong. The Polish armed revolt

---

[1] Anna Carlotta Leffler, Duchess of Cajanello (1895). *Sonya Kovalevsky*, transl. by A. de Furuhjelm. T. Fisher Unwin.

[2] In a hint of one of the public health revolutions to come, the smallpox vaccine had been invented.

The Women of the Moon. Daniel R. Altschuler Stern and Fernando J. Ballesteros Roselló.
© Daniel R. Altschuler Stern and Fernando J. Ballesteros Roselló 2019. Published in 2019 by Oxford University Press. DOI: 10.1093/oso/9780198844419.001.0001

in 1863 against the Russian domination lasted two years, although it was impossible to triumph against the imperial forces. Thousands of Poles lost their lives, and afterwards many emigrated to seek a better life, as we shall learn when we turn to Marie Skłodowska. Meanwhile, in Russia, a new progressive intellectual class was born, "the nihilists," opposed, as always happens, to the most conservative elites. They thought that society had to be reformed, that through science there would be a better future, that women had the same rights as men, and that it was necessary to educate the peasants. They called themselves "children of the 60s."

This was the world into which, on January 15, 1850, Sofia Krukovskaya[3] (also Sophie, Sofie, Sofya, Sonia, Sonja, and Sonya) was born, second daughter of a lieutenant colonel of the Russian Imperial Army (under the command of Tsar Nicholas I), Vasily Vasilyevich Korvin-Krukovsky, and Yelizaveta Fedorovna Schubert, a woman from a family of German-Russian intellectuals, granddaughter of a distinguished astronomer, Theodor Schubert (1758–1825), and daughter of a prominent land surveyor, Fiador Fiodorovich Schubert (1789–1865).

Nineteenth-century Russia was dominated by men; a woman's role was to raise children, maintain a home, and serve her husband. The "proper" Russian woman learned to speak French, to play a musical instrument, to paint, and to converse with the purpose of attracting a good husband. It was accepted that women were intellectually inferior and it was inconceivable that they would study at a university, a redoubt of men.

Sofia's interest in mathematics awoke at an early age. She tells us in an autobiographical essay of her childhood bedroom: in 1858 her family moved from Moscow to an estate in Palibino, south of St. Petersburg, near the Polish border[4]. The walls of her room there were wallpapered with notebook pages from a course in integral and differential calculus given by professor Mikhail Vasilyevich Ostrogradsky (1801–1862). Although she did not understand the content, those strange pages on the wall aroused her scientific curiosity and, she writes, left in her mind a deep trace that facilitated her understanding of mathematical analysis. She

[3] Ann Hibner Koblitz (1933). *A convergence of lives: Sofia Kovalevskaia. Scientist, writer, revolutionary*, Rutgers University Press.

[4] Michele Audin (2011). *Remembering Sofya Kovalevskaya*, p. 29. Springer.

said that she spent hours trying to understand the order that the disorganized pages stuck on the wall should follow.

As the child of a wealthy family, Sofia was educated by a tutor who taught her history, literature, and mathematics. Her uncles, Fedor Fedorovich Schubert and Pyotr Vasilevich Korvin-Krukovsky, spoke to her about science and mathematics, encouraging her interest. Her talent was evident: at the age of fourteen she derived certain trigonometric relations on her own to understand a chapter on optics in a physics book written by Nikolai Nikanorovich Tyrtov (1822–1888), professor of physics at the St. Petersburg Naval Academy and a neighbor of the family. Tyrtov, initially incredulous, was impressed by Sofia's ability and persuaded her father, who initially was skeptical of the idea, to allow her to study calculus and trigonometry with Aleksandr Strannolyubsky (1839–1903), professor at the naval academy and advocate for the education of women and workers, who quickly recognized the young woman's mathematical ability.

Exhausting the academic opportunities in St. Petersburg, Sofia wanted to continue university studies, but universities in Russia (as in most of Europe) did not admit women. To travel to another country, with conditions less unfavorable for a woman, she needed to be accompanied by her father or husband. So, following the advice of her older sister and feminist activist (and friend of Fyodor Dostoyevski, incidentally), Aniuta (1843–1887), she agreed to marry a friend, Vladimir Onufrievich Kovalevsky (1842–1883).

This move was quite common at the time; fictitious marriages were regularly used to escape the tyranny exercised by men over women. Her father was initially opposed, but in the face of her determination to study, he ultimately backed down and accepted the arrangement. Vladimir was a publisher of scientific and political books, a business that was not going too well. As a young idealist who wanted to help the talented Sofia, he understood that "at first men should accept second place, to give women time to outgrow their past servitude"[5].

After marrying in 1868, Sofia adopted, as was customary, the feminized surname of her husband: Kovalévskaya. In 1869, after a time in St. Petersburg, the Kovalevsky couple made their way to the University of Heidelberg, in Germany, where Vladimir began studying geology, but

---

[5] Ann Hibner Koblitz (1993). *A convergence of lives: Sofia Kovalevskaia. Scientist, writer, revolutionary*, p. 83. Rutgers University Press.

Sofia's request to enroll was denied. However, a special committee convened to consider her situation determined that each professor could decide whether to allow Sofia to attend—unofficially—his classes. Some refused to teach if there was a woman in the room, but the remainder made the move more than worth it: the mathematician Leo Königsberger (1837–1921) admitted her as a disciple, and she studied physics with the eminent Kirchhoff (Gustav Robert Kirchhoff 1824–1887) and physiology with Helmholtz (Hermann Ludwig Ferdinand von Helmholtz 1821–1894). Word of the great mathematical ability of the young Russian quickly spread. After a year of study—and a year of economic difficulties, conflicts between formal studies and her feelings (her marriage to Vladimir was not fictional enough for her not to care, or real enough to consummate it), as well as conflict within the small émigré community in Heidelberg—she determined that she wanted to study with the eminent Karl Weierstrass (1815–1897) in Berlin, one of the most distinguished mathematicians of the time (a small crater on the Moon is named after him).

Sofia and Vladimir, who had traveled to Jena to continue his studies in geology, separated. Their fictitious marriage was in difficulties and Vladimir was considering divorce. In light of this situation, Sofia traveled with her friend Julia Lermontova (1847–1919), a chemistry student and the first woman to receive a doctorate, to Berlin in 1870.

Although she presented herself with a recommendation from Weierstrass's disciple Königsberger, he received Kovalévskaya with some reserve. He decided to put her to the test, and gave her some problems that he had prepared for his more advanced students, convinced that she would not be able to solve them. Kovalévskaya showed up a week later to tell him that she had solved the problems. Weierstrass, still skeptical, sat down next to her to see what she had done. To his surprise not only were the solutions correct, but they showed great ingenuity. The professor was delighted, and from that moment a relationship of parental love between the great mathematician and the young Russian was established.

Weierstrass, with the support of other professors—Hermann von Helmholtz, Rudolf Virchow, and Emil du Bois-Reymond—petitioned the university to allow Sofia's attendance at their classes. The academic senate refused, so Weierstrass accepted her as a private student.

In 1874, after four years of study, Weierstrass got the prestigious University of Göttingen to consider three papers by Kovalévskaya (one on the geometry of Saturn's rings, another on Abel integrals, and a third on partial differential equations). On the basis of these, the university granted her a doctorate in absentia, summa cum laude, at twenty-four years of age, without need for a defense or an oral examination, an unheard-of attestation to their quality.

She became the first woman outside of Italy to obtain a university doctoral degree in mathematics. (The first woman to receive a doctorate from a university was the philosopher Elena Lucrezia Cornaro Piscopia (1646–1684) from the University of Padua, in the year 1678.)

Despite this success on the academic career path, she did not get a job. After fights and reconciliations, the Kovalevsky couple thought that they might obtain academic work in St. Petersburg (Vladimir had completed his studies in Jena), and that they could transform their fictitious relation into a real one. Although they were well received in St. Petersburg, things did not go as they imagined and Sofia dedicated her time to literature and political activism, while Vladimir undertook financial investments aiming at economic stability without depending on the labor market. Her father, Vasily Vasilyevich, died in 1875 and left behind a non-negligible inheritance that the couple used to invest, giving life to this possibility. Their marriage of convenience became a real one: in 1878, their daughter Sofia Vladimirovna (nicknamed Fufa) was born. But this personal success was short-lived; they failed economically and ended in ruin, moving to Moscow in an attempt to rebuild their lives.

But Sofia's passion for mathematics called on her. At a congress in St. Petersburg in 1879, she presented work that was well received not only by Russian mathematicians but also by a Swedish mathematician and disciple of Weierstrass, Magnus Gustaf (Gösta) Mittag-Leffler (1846–1927) (who plays a role in the stories of other women of the Moon). Wishing to return to work in mathematics, she returned to

Figure 45 Gustaf (Gösta) Mittag-Leffler.

Berlin in 1880 to work with Weierstrass, an interlude that served for Sofia to restart her mathematical career. Still searching for stability, she moved to Paris for a season, along with her sister Aniuta (now Aniuta Jaclard, having married the French socialist revolutionary Charles Victor Jaclard (1840–1903), member of the First International and the Paris Commune), and became involved with revolutionary groups. At the same time, she managed to continue with her mathematical work, and met Charles Hermite (1822–1901), Émile Picard (1856–1941), and the illustrious Jules Henri Poincaré (1854–1912), and completed a study on the refraction of light. She was elected to the Mathematical Society of Paris in 1882.

Her relationship with Vladimir was increasingly unstable, and coinciding with the assassination of Tsar Alexander II in early 1882, the Kovalevskys dissolved their marriage, she at the age of thirty-one. Sofia returned to Berlin with Fufa, and Vladimir traveled to Odessa to see his brother. Overwhelmed by his financial and personal problems, Vladimir committed suicide by drinking a bottle of chloroform on the night of April 23, 1883. In a letter to his brother he wrote[6]: "Write to Sofa that my constant thoughts were of her and of how much I am at fault before her, and how I spoiled her life, which without me would have been bright and happy. My last request to Aniuta is that she look after her and little Fufa, she (Aniuta) is the only one now capable of doing that, and I beseech her to do so." The news caused Sofia great anguish, to the point of needing medical intervention. She felt guilty that she had not taken enough care of Vladimir when it was evident that he had serious problems—that more than her husband he had been her friend.

Despite this, Sofia's career was starting to gain traction. In 1883, Mittag-Leffler, defender of women's rights and professor of mathematics at the new University of Stockholm, arranged for her a probationary appointment for one year and without salary (Privatdozent), an offer that Sofia accepted, since she had no choice. The appointment caused controversy; the Swedish writer Johan August Strindberg reflected the thinking of many at that time when he wrote in a Stockholm paper: "A female teacher is a pernicious and unpleasant phenomenon, one might even say, a monstrosity."

In January 1884, she gave her first class on differential equations to a group of twelve enrolled students. But the room was filled with other

---

[6] Ann Hibner Koblitz (193). *A convergence of lives: Sofia Kovalevskaia. Scientist, writer, revolutionary*, p. 171. Rutgers University Press.

students, professors, members of the press, and curious citizens, aware of the historical moment. Professor Kovalévskaya began her class nervously, but it ended to the applause of those present. Despite her loneliness, the alien social environment, the prejudice of many colleagues—against her gender, her foreign origin, and her nihilist politics—and the foreign language, she was able to perform successfully. After her probationary year, Kovalévskaya, thanks in part to the maneuvers of Mittag-Leffler, was finally appointed as a professor, the first woman appointed to such position in a northern university. (There is a precedent in Italy: Maria Gaetana Agnesi (1718–1799), the first mathematician appointed to a chair at the University of Bologna by Pope Benedict XIV.)

Gösta Mittag-Leffler also appointed her as the first woman editor of a professional journal, the prestigious *Acta Mathematica*, which he had founded in 1882. In this new role she nurtured a fruitful correspondence with mathematicians throughout Europe, particularly in France (Hermite), Germany (Weierstrass and Kronecker), and Russia (Chebyshev), and in her travels she was able to establish warm relationships, both personal and professional, with many of them, without which a scientist, with rare exceptions, cannot be productive.

Gradually, some barriers came down. Her career began to be noticed in the press, and in 1884, the University of Berlin's rector, who had previously forbidden her to attend lectures because she was a woman (a position that seems incredible at the distance of a century and a half), issued a decision allowing Sofia to attend any conference that took place at any university in Prussia—although she was the only woman with that privilege. In a letter to her friend Maria Jankowska-Mendelson (member of the revolutionary proletarian Polish party) from January 1884, she told her about a new project related to the rotation of a solid body[7]:

> The new mathematical work that I have started interests me intensely and I would hate to die without discovering that which I am looking for. If I succeed in solving the problem, my name will integrate the list of the most prominent mathematicians. According to my calculations I need another five years to get good results. But I hope that in five years there

---

[7]   Ann Hibner Koblitz (1993). *A convergence of lives: Sofia Kovalevskaia. Scientist, writer, revolutionary*, p. 186. Rutgers University Press.

will be more than one woman capable of taking my place here, and I can devote myself then to other urges of my gypsy nature.

By the end of 1886, her beloved sister Aniuta was hospitalized in Russia with cancer. Sofia traveled several times to be by her side. The family arranged to move her to Paris, for better medical treatment and to be near her husband. The sudden news of Aniuta's death in 1887 saddened Sofia deeply, much more deeply than Vladimir's suicide. Sofia would write: "I only cry and I mourn when I am a little sad. When I feel great anguish, I remain silent. No one can detect my suffering."

During this period, she met Alfred Nobel (1833–1896) and also the Norwegian polar explorer Fridtjof Nansen (1861–1930), with whom she maintained an ephemeral romantic relationship. She also met Maksim Maksimovich Kovalevsky (1851–1916), Vladimir's cousin and recently professor of law at Moscow University, dismissed in 1887 for his liberal politics. A turbulent romance began between Sofia and Maksim, who settled in Paris, a city to which Sofia traveled frequently. She could live neither with him nor without him.

Christmas Eve of 1888 was a high point for Sofia's career, as she was awarded the prestigious Bordin prize from the French Academy of Sciences. Encouraged by Mittag-Leffler and Weierstrass, she had submitted a monograph for the contest: *Mémoire sur un cas particulier du problème de le rotation d'un corps pesant autour d'un point fixe*, solving a difficult physical-mathematical problem related to the rotation of a solid body, at which she had worked for five years, as she had estimated it would take her in her letter to Jankowska-Mendelson. Her desire to solve it before she died was fulfilled.

The rules of the competition specified that the work be submitted anonymously in an envelope, identified with a legend, accompanied by another sealed envelope containing the author's name and the same legend; this second envelope would not be opened until the jury issued its decision on the basis of the work alone. In 1888, the jury was so impressed with the winning work (of fifteen submitted) that it decided to increase the prize of three thousand to five thousand francs. Surely more than one member was surprised to open the second envelope and discover that the winner was a woman. Sofia's legend was: "Say what you know, do what you must, no matter what happens." (*Die ce que tu sais, fais ce que tu dois, adviendra que pourra!*).

She was not the first woman awarded a prize by the academy; another mathematician, another Sofia even, had preceded her in 1816, the Frenchwoman Sophie Germain (1776–1831), who also had her "Weierstrass" in the figure of the illustrious French mathematician Joseph- Louis Lagrange (1736–1813). Sophie Germain also has her crater, but it is on the surface of Venus.

In his speech on the occasion of the presentation of Kovalévskaya's award, the president of the academy said: "The members of the jury have determined that the work bears witness not only to a broad and deep knowledge, but also to a mind of great imagination." Many banquets were held in her honor and the European press reported her triumph.

Weierstrass wrote to her: "I do not need to tell you how much I have rejoiced with your success, as well as my sisters and all your friends here. In particular, I have experienced a real satisfaction, because competent judges have now pronounced their verdict that my faithful student, my 'weakness', is not a frivolous puppet."

After the great effort to complete her work, Sofia felt exhausted and depressed, a feeling she had experienced to a lesser degree after receiving her doctorate and also after obtaining her first university position in Stockholm. To Mittag-Leffler she wrote: "I receive so many letters of congratulations and as if it were an irony of fate, I have never felt so miserable in my life."

After this recognition she was offered (not without the need for a campaign in her support from Mittag-Leffler and other friendly mathematicians) a lifelong chair at the University of Stockholm, becoming the first woman so appointed, and in 1889, after the statutes were changed, she was elected as the first woman correspondent[8] member of the Russian Imperial Academy of Sciences (although one must note that correspondent members did not have the same luster as proper members).

In February of 1891, back in Stockholm from a trip through southern France with Maksim, during which they had decided to marry, Sofia contracted the flu. On her trip she passed through Paris and Berlin and met with several mathematicians, including her beloved Weierstrass, and on her return to Stockholm, despite not feeling well, she gave her class and went to a meeting to raise funds for the university. At night she

---

[8] A category established in some countries to designate members who were abroad and communicated with the academy by correspondence.

felt worse and the next day she requested a doctor. Nurses attended her at home, but her condition worsened at night, and she fell into a coma from which she never recovered. Pneumonia took her life at the age of forty-one. She was buried in Stockholm, and her tomb was decorated with a wreath of white lilies from her dearest friend: Weierstrass. Those who knew her agreed: she was a brilliant, professional, and tenacious woman, who in forty-one years lived what others would not live in eighty-two. In her biography, her friend Carlotta Leffler wrote[9]:

> Sonya's life had been longer than most. She had lived intensely; she had drained the cup both of sorrow and of joy. She had quenched the thirst of her spirit at the wells of wisdom. She had risen to the heights to which genius and imagination alone can carry the soul. To others she had given instinctively of her knowledge, experience, fantasy, and feeling. She had spoken with the inspiring voice which genius alone possesses when it does not isolate itself in selfish retirement. No one who knew her could remain unmoved by the influence ever exercised by the keen intellect and glowing feeling, which spread sunshine and growth around. Her mind was fertile because her intellect was unselfish. Her highest aspiration was to live in mental union with another.

Asteroid 1859 Kovalevskaya was discovered in 1972 by L.V. Zhuravleva, and named in her honor. It is located in the asteroid belt, between Mars and Jupiter, and has a diameter of 46 km. You won't be able to see crater Kovalevskaya, since it is on the far side of the Moon. It is one of the largest of our women's craters, at 115 km in diameter, with a mound in the center, and beautiful stepped walls.

---

[9] Anna Carlotta Leffler, Duchess of Cajanello (1895). *Sonya Kovalevsky*, transl. by A. de Furuhjelm. T. Fisher Unwin.

Figure 46 Location of crater Kovalevskaya. Courtesy of the Lunar and Planetary Institute, Houston, Texas.

Figure 47 Lunar Reconnaissance Orbiter zoom on crater Kovalevskaya (image width is 150 miles).

11

Annie Scott Dill Russell Maunder (1868–1947)

Figure 48 Annie Scott Dill Maunder (née Russell) by Lafayette (Lafayette Ltd) half-plate nitrate negative, December 7, 1931 NPG x47933. National Portrait Gallery, London.

The large volume of her historical work put Mrs. Maunder ahead of her time in the now popular field of ancient and archaic astronomy. In her lifetime she was looked upon as an expert on this subject: like her eclipse observations, her efforts may not have been sufficiently appreciated by later generations.

MARY T. BRÜCK[1]

Annie Scott Dill Russell was born in Ireland, daughter of the Reverend William Russell (1824–1899), minister of the Presbyterian Church in Strabane, and his second wife, Hessy Nesbitt Dill. As a teen, she studied at the Ladies Collegiate School in Belfast (later Victoria College), and after winning a prize for academic excellence, entered at the age of eighteen Girton College of the University of Cambridge, where in 1889 she graduated with honors in mathematics[2]. She did not receive a formal diploma since this was only granted to men (this was changed in 1928).

In 1890, she began work at the Greenwich Royal Observatory to work as an assistant to the solar astronomer Walter Maunder, along with

[1] Mary Brück (2009). *Women in early British and Irish astronomy: stars and satellites*, p. 230. Springer.
[2] Mary T. Brück (1994). Alice Everett and Annie Russell Maunder: torch bearing women astronomers. *Irish Astronomical Journal*, Vol. 21 (3/4), p. 281.

The Women of the Moon. Daniel R. Altschuler Stern and Fernando J. Ballesteros Roselló.
© Daniel R. Altschuler Stern and Fernando J. Ballesteros Roselló 2019. Published in 2019 by Oxford University Press. DOI: 10.1093/oso/9780198844419.001.0001

other women "computers," forming part of the first group of female salaried astronomers in Great Britain. It was the royal astronomer Sir William Christie who first employed women in the observatory as computers[3]—skilled and inexpensive labor. Before the advent of mechanical and then electronic calculators and computers, calculations to solve complex problems were carried out by human computers, who worked on a sort of intellectual assembly line, building up complex calculations from their constituent parts to obtain results. Human computers carried out their boring and systematic work in an environment described by Charles Dickens (1812–1870) in his novel *Hard Times*[4]: "a stern room, with a deadly statistical clock in it, which measured every second with a beat like a rap upon a coffin-lid." It was not until after the Second World War that human computers turned their work over to machines, and acquired new jobs as operators and programmers of the machines.

Walter Maunder is now best remembered for his retrospective observation that between 1645 and 1715 almost no sunspots were observed, a period known as the Maunder minimum (possibly associated with a decrease in global temperature called the Little Ice Age, which coincided at the time with this decrease in solar activity).

He also concerned himself with the social institutions in which astronomy took place: he founded the British Astronomical Association (BAA) in 1890 in protest at the exclusivity of the Royal Astronomical Society—both its high membership costs and its exclusion of female members. Walter had been widowed in 1888 (his wife, Edith Hannah Bustin (1852–1888) died of tuberculosis) and married Annie in 1895. According to civil service employment regulations, when he married Annie, she had to resign her position, but she continued to work with Walter on solar research and as editor of the BAA journal, a position she held for fifteen years. Annie also devoted herself to solar photography and historical studies of astronomy.

Astronomy in the nineteenth century was a global affair, and like many other women of the moon, the Maunders traveled the globe to make observations. The year 1898 saw the couple in India to observe a total eclipse

---

[3] Edward Pickering would follow the same practice at Harvard College Observatory in the United States, and we shall meet several of the women who built the basis of modern astrophysics later in this book.

[4] Charles Dickens (1864). *Hard times*, Chapter XV.

Figure 49 The coronal streames. Drawing by Mr. W. H. Wesley from photographys by Mrs. Maunder.

of the Sun on January 22. Annie had built a special wide-angle camera with which she photographed the solar corona. Our Agnes Clerke, after seeing the results of several expeditions, pronounced[5]: "Regarding the corona, Mrs. Maunder, with her small lens, has won over all the great instruments." (In the lower right part of the image the planet Venus was observed.)

Not just a global but a family pursuit for the Maunders, in 1900, Walter brought his daughters (from his previous marriage) Irene (age twenty) and Edith (age twenty-two) along on an expedition to observe another total eclipse of the Sun, this time from Algeria. The family installed themselves (and their equipment) on the roof of the hotel "de la Regence." In the report written by Walter Maunder we read that they were not mere tourists, but participants in the scientific work as well[6]:

Another observer, Miss Irene Maunder, describes the effect of totality: A bell rang and we all hurried to our places, for we knew there were but five minutes to totality. Another bell, but one minute more. The sky was deep purple, while over the sea was a strange light on the horizon, a compromise

[5] In a letter to David Hill in 1898.
[6] Edward Walter Maunder (ed.). *The total solar eclipse, 1900. Report of the expeditions organized by the British Astronomical Association to observe the total solar eclipse of 1900, May 28*, p. 74. British Astronomical Association. https://ia800202.us.archive.org/7/items/totalsolareclips00maunuoft/totalsolareclips00maunuoft.pdf

between a thunderstorm and a sunset. The colour faded from, the sea and trees, a shouting and wailing arose from the square below, the light was fading; suddenly the moon slipped over the sun and the eclipse was total. "Go!" shouted a loud voice; a metronome began to beat seconds, and as its bell rang at each sixth stroke, my sister called the time. "One! Two! Three! Four! Five! Six!" There! My photographs were taken, and now I could look up! I shall never forget the sight. A deep purple sky, a black globe, sur-rounded by a crimson glow, and above and below it a milk-like flame stretching its long streamers away into the purple. The darkness, the cold wind, the silent workers around me, and the shouting crowd below all tended to make this strange and glorious sight still more impressive, and I found myself stretching out my arms to that exquisite corona in a perfect ecstasy.

In 1904, a famous diagram was published under Walter's name, but we know now that it was a joint work. Plotting the latitude of sunspots over time, it shows that sunspots move from high latitudes (north and south) to the solar equator with the passage of time during each solar cycle. It is called the butterfly diagram[7].

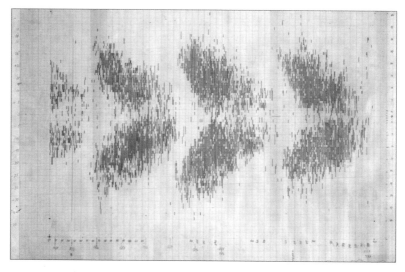

Figure 50 The original butterfly diagram (1904) showing the equatorward migration of sunspots as the solar cycle progresses.

[7] The original butterfly diagram is now under the care of the High Altitude Observatory. During the Second World War, Annie Maunder passed on this piece of solar

In 1910, the Maunders published a book, *The Heavens and Their Story*. In the preface, Walter writes that the book, "whose authors are the names of my wife and mine, is almost entirely the work of my wife, since circumstances prevented me from taking part in her writing soon after it began."

With the coming of the First World War, many of the observatory's staff were enlisted and Walter, already retired, returned to continue with solar work; Annie joined him as a volunteer without pay. In 1916, she was the first woman elected to the Royal Astronomical Society, after the statutes were changed to accept women (she had been rejected when she was nominated in 1892).

Walter died in 1928, and Annie continued to dedicate herself to the work of the BAA and to historical studies of ancient and archaic astronomy. She contributed to a history of the BAA (initially written by Mary Evershed[8]) and in 1942, at age seventy-four, presented a paper recalling the first meeting of the BAA[9]: "Men and Women astronomers came in on equal terms: so also, the rich and the poor, those who worked with their hands and those with their heads; and all pooled their varying knowledge for the public good."

Her work about the origin of the Western constellations, which she estimated to date from the year 2900 BCE, is noteworthy, and agrees with more recent studies[10]. She died in 1947 at the age of eighty. The Maunder crater, in honor of both, is a large 55 km crater on the far side, inside the Mare Orientale. Like Kovalevskaya, it is crowned by a central mound and its walls are staggered.

research legacy to Walter Orr Roberts (the first director of High Altitude Observatory) for safe keeping. The eclipse drawing is also there.

    [8] The History of the British Astronomical Association: The First Fifty Years (1947). *Memoirs of the British Astronomical Association*, Vol. 36, pp. 1–132.

    [9] A.S.D. Maunder (1943). Reminiscences of the British Astronomical Association. *Journal of the British Astronomical Association*, Vol. 42, p. 268.

    [10] John Rogers (1998). Origins of the ancient constellations: I and II. *Journal of the British Astronomical Association*, Vol. 108, no. 1, pp. 9–28, and no. 2, pp. 79–89.

Figure 51 Location of crater Maunder. Courtesy of the Lunar and Planetary Institute, Houston, Texas.

Figure 52 Lunar Reconnaissance Orbiter zoom on crater Maunder (image width is 150 miles).

12

Williamina Paton Fleming
(1857–1911)

If one could only go on and on with original work, looking for new stars, variables, classifying spectra and studying their peculiarities and changes, life would be a most beautiful dream; but you come down to its realities when you have to put all that is most interesting to you aside, in order to use most of your available time preparing the work of others for publication. However, "Whatsoever thou puttest thy hand to, do it well." I am more than contented to have such excellent opportunities for work in so many directions, and proud to be considered of any assistance to such a thoroughly capable Scientific man as our Director (Edward Pickering).

From the diary of WILLIAMINA FLEMING, March 5, 1900

Figure 53 Williamina Fleming. Portrait courtesy of Special Collections, Fine Arts Library/ Harvard University.

Williamina Fleming, born in Scotland in 1857, was the first of the "Harvard computers," a group of female astronomers who, based at Harvard University, revolutionized astronomy in the early twentieth century. The origins of this productive group lay, surprisingly, in the hobby of one doctor Henry Draper (1837–1882), Professor of Physiology and Chemistry at the University of New York, amateur astronomer, and one of the pioneers in obtaining stellar spectra and excellent images of the Moon.

If a crystal prism is illuminated with a beam of white light, the light breaks down into the colors of the rainbow. This is what is known as the light spectrum, with the different colors of light corresponding to

The Women of the Moon. Daniel R. Altschuler Stern and Fernando J. Ballesteros Roselló.
© Daniel R. Altschuler Stern and Fernando J. Ballesteros Roselló 2019. Published in 2019 by Oxford University Press. DOI: 10.1093/oso/9780198844419.001.0001

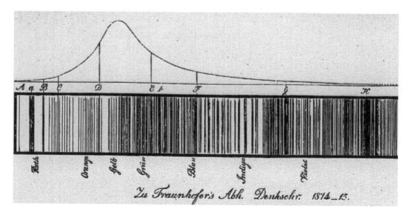

Figure 54 Drawing of the solar spectrum by Fraunhofer.

different wavelengths (or frequencies). Sunlight, when broken into its constituent colors in this fashion, does not present a smooth gradient with one color fading into the next (as does the rainbow); superposed on the colors one finds narrow dark bands These features in the Sun's light were first observed by the English chemist William Wollaston (1766–1828) and studied in detail by the German physicist Joseph von Fraunhofer (1787–1826). A drawing[1] of his observations of 1817 is shown in Figure 54.

In 1849, the German physicist Gustav Kirchhoff (1824–1887) discovered that the chemical composition of a gas can be deduced from its luminous spectrum, no matter how far the observer is from the light source, thanks to the bright and dark lines (respectively emission and absorption lines) that appear in this spectrum; each chemical has its own characteristic pattern of lines, in the manner of a fingerprint.

The possibility of knowing at a distance the chemical composition of the faraway stars (and also their temperature) by analyzing only the spectrum of their light, implied a marvelous outlook to the astronomers of the time. At the same time, it put an end to a pessimistic (and misguided) prediction about the limits of science, that of the French philosopher August Comte (1798–1857), who in 1835 stated that: "As regards the stars, although we can conceive the possibility of determining their forms, their sizes and their movements, we will never be able by any means to study their chemical composition."

---

[1] *Denkschriften der Königlichen Akademie der Wissenschaften zu München*, 1814–15, pp. 193–226.

Henry Draper, seeing the power of this tool, became an expert in the acquisition of stellar spectra. He took the first of them in 1872, recording the star Vega, the fifth-brightest star in the sky at a distance of twenty-five light-years, on a photographic plate, and continued to record spectra for the next ten years, until 1882. That year, Draper went on a hunt in the Rocky Mountains, from which he returned with a bad cold. Despite this, on his return to New York, the Draper couple invited the members of the National Academy of Sciences to dinner, event for which Draper had installed several electric light bulbs of Mr. Edison as lighting for the dinner table. Among the guests was astronomer Edward C. Pickering, director of the Astronomical Observatory at Harvard University, who expressed a keen interest in the stellar spectra Draper had obtained (more than a hundred), and offered to take measurements of them if they could be sent to the Observatory. Nobody that evening suspected that he would not see Draper again, because after dinner the cold turned into pneumonia that took his life within days. The rich legacy of his star spectra, which contained valuable information waiting for someone to analyze it, was left to the care of his widow, Mary Anne.

For his part, Pickering began to enhance and develop the spectroscopic work at the Harvard Observatory, continuing in a certain way Henry Draper's work. He showed his spectroscopic works to Draper's widow, who had long wanted to establish some kind of memorial tribute to the work of her late husband. Knowing that the Harvard observatory was short of funds, after some negotiations she decided to finance Pickering's research in exchange for the establishment of the Henry Draper Memorial, to continue the work of her husband[2]. Pickering accepted, and the Harvard Observatory set out to analyze and classify Henry Draper's plates, as well as to take more stellar spectra to add to the collection. The goal was ultimately to extend the work initiated by Draper to all light-producing objects in the sky. Initially the program was limited to the sky accessible from Harvard, but in order to observe the stars of the Southern Hemisphere, in 1890 the Harvard Observatory established an observation station in Arequipa, Peru. As Pickering wrote, it was expected that "the final results would constitute a complete study of the constitution and condition of the stars."

---

[2] Dava Sobel (2016). *The glass universe*. Viking.

But let's go back a few years and introduce the protagonist of this story, Williamina Paton Fleming. As mentioned, she was born in Scotland in 1857. Her father (an amateur photographer, in a foreshadow of the photography that would be an important part of Williamina's life) died when she was 6 years old, leaving the family destitute. Williamina entered the working world as an apprentice school teacher at age 14, doing paid internships (income that suited the family very well), and she proved to be a very capable student. But she abandoned these studies in 1877 when, at 20, she married the banking accountant James Orr Fleming. The following year they would emigrate to the United States, taking up residence in Boston; but a few months after arriving, Williamina became pregnant and her husband abandoned her[3].

Alone, in a foreign country, with no income and waiting for a child to be born, little did she suspect that her misfortune would be her doorway to immortality (at least as we astronomers understand it). In need of money, she went to work as a housekeeper at the home of Edward Pickering, director of the Harvard Observatory.[4] Apparently, shortly after her arrival, Pickering argued heatedly at the Observatory with one of his assistants, disappointed by the poor quality of the work he had done. Angry, he snapped, "Even my Scottish housekeeper could do a better job!" Backing up his words with deeds, he put Williamina (then twenty-four years old) to do the job of his assistant. Williamina proved him right: despite not having a scientific background, she did the job much better than Pickering's assistant. As a result, he hired her as a part-time assistant at the observatory, while continuing with her duties as a housekeeper.

This lasted only a few months. As her delivery date approached, Williamina decided to return to Scotland to be with her family and receive help from her mother. Pickering, a brilliant man, evidently impressed her, for Williamina baptized her son Edward Charles Pickering Fleming. She remained in Scotland for a year and a half, and nothing seemed to indicate that she was going to return to Boston; but her work at the Observatory and her new life there must have pleased her, because she decided to go back to Harvard. She left her son Edward in the care of her mother and returned to her part-time job as a housekeeper and computer. Pickering

---

[3] There is no record of what happened to her husband after this, although it is believed that he died before 1900, since Williamina appears as a widow in the United States census of that year.

[4] Although this is probably just an amusing legend.

Figure 55 Computers in 1891. Williamina Fleming (standing center), Antonia Maury (third woman form left), Annie Cannon (in the center with magnifying glass), with Pickering (the one with a beard). Harvard Archives HUV 1210, 9-6.

must have been impressed in turn with Williamina's work, since shortly after her return to Harvard he went on to hire her at the Observatory full-time. Williamina would not see her son until six years later: in 1887, Edward, by then eight years old, traveled to Boston with his grandmother and two cousins; none of them would return to Scotland.

When the observatory received its windfall of funding from Draper's widow, the year was 1886 and Williamina had been on the observatory's staff for 5 years, leading a small group of women computers. It soon became evident that the current staff of the observatory were not going to be able to cope with all the spectra that the project of the Henry Draper Catalogue was producing. More staff were needed. Pickering felt that not only were women more patient than men for routine work, but they had a greater capacity for observation. This made them ideal for the analysis of spectra on photographic plates. Given the good results from hiring Williamina, Pickering had opened the observatory to women. Ever the canny manager, he was also keenly aware that women could be hired for less pay than men, and therefore he could bring on a larger team to analyze such a large amount of data than if he hired men (in fact, some women even offered to work as volunteers without pay in order to work in science).

This female team was directed by Williamina, who was also in charge of selecting and hiring them. They were collectively called the "Harvard computers" because they performed the tedious classification of the spectra and the analysis of the position and intensity of the stars from the photographic plates. This term, which today sounds so informatic, was already in use since the seventeenth century with the meaning of "person who computes," performing mathematical calculations following fixed rules without deviating from them. However, this group of women was also known by a more malicious nickname: "Pickering's Harem."

The first Henry Draper Catalogue was published in 1890 with data from more than 10,000 stars. This demonstrated the power of an assemblage of people analyzing data even if they did not have specialized scientific training. This practice continues today: numerous projects with a huge amount of data to analyze recruit volunteers to help them. Stardust@home is a prime example, a project to analyze data from the Stardust space mission (which collected dust from a comet on an aerogel substrate). Stardust@home made available to the public 40 million scanned images of the aerogel so that human eyes (still better at recognizing patterns than any computer program) could find traces of dust. Stardust had 23,000 volunteers, and the Clickworkers project to identify and visually count craters on Mars was even more popular, with 85,000 volunteers. Or the Galaxy Zoo project, to classify galaxies by their appearance from images of the Sloan Digital Sky Survey, with more than 200,000 users. Today there are several dozen "citizen science" projects in operation.

Parceling out the work in a crowdsourced project today is, of course, facilitated by computer databases, but 120 years ago, this was a managerial task. Similarly, tracking and cataloguing raw data was part of project management for Williamina. In addition to her job as "head of computers," she was also in charge of indexing the photographic plates where the spectra were taken, so that they would be easy to find when needed (it is estimated that throughout her life she would examine about 200,000 photographic plates). To make this indexing possible, and to order the different types of spectra, Williamina and Pickering developed a classification system, which ordered the stars by the intensity of their hydrogen absorption line and the presence of other lines, following the letters of the alphabet (A, B, C, ... up to 22 types). But this classification, which was very useful as an ordering system, did not seem to have any clear physical meaning. It would fall to another of the Harvard computers—another woman of the Moon—Annie Jump Cannon, with whom Williamina maintained a good friendship, to find the order that underlay the stellar spectra.

In 1899 and after 12 years as de facto manager of the archive, Williamina Fleming officially became the person in charge of the photographic archive (Harvard's Curator of Astronomical Photographs), the first corporate position of Harvard held by a woman.

Pickering had his lights and his shadows. On the one hand, he was sympathetic to women's suffrage, as well as a strong advocate for women's access to the university. Without implying impropriety, it has sometimes been said that the name "Pickering's Harem" was not entirely incorrect: the term harem derives from the Arabic *harim*, meaning "forbidden place," and Pickering had indeed allowed the entrance of women to a forbidden place, the observatory of a male university. On the other hand, he did not encourage the women of the Observatory to pursue their own investigations—from his point of view, the main objective of the computers should be the analysis of the spectra and the Henry Draper Catalogue. Of this, Williamina complained in her diary[5]:

> Looking after the numerous pieces of routine work which have to be kept progressing, searching for confirmation of objects discovered elsewhere, attending to scientific correspondence, getting material in form for publication, etc., has consumed so much of my time during the past few years that little is left for the particular investigations in which I am especially interested. The Director, however, says that my time employed in the above work is of more value to the Observatory so I have delegated my measures of variables, etc., to Miss Leland and Miss Breslin. I hope, however, to be able soon to finish the measures of the "out of focus" plates, and to get well settled down to my general classification of faint spectra for the New Draper Catalogue.

Pickering himself was aware of this, for on one occasion he wrote that Williamina's tasks "occupied so much of her time that they seriously interfered with her scientific investigations."

---

[5] This and other quotes are taken from the Harvard University Archives. *Chest of 1900: Diaries, Journal of Williamina Paton Fleming.*

Pickering also believed that he paid his women more than enough, which they did not agree with at all. Williamina wrote in her diary:

> I had some conversation with the Director regarding women's salaries. He seems to think that no work is too much or too hard for me, no matter what the responsibility or how long the hours. But let me raise the question of salary and I am immediately told that I receive an excellent salary as women's salaries stand. If he would only take some step to find out how much he is mistaken in regard to this he would learn a few facts that would open his eyes and set him thinking. Sometimes I feel tempted to give up and let him try someone else, or some of the men to do my work, in order to have him find out what he is getting for $1500 a year from me, compared with $2500 from some of the other assistants. Does he ever think that I have a home to keep and a family to take care of as well as the men? But I suppose a woman has no claim to such comforts. And this is considered an enlightened age!!.

Despite the crush of everyday tasks, Williamina did manage to take time to do her own research. She discovered fifty-nine nebulas, some of them very famous among lovers of astrophotography, such as the "Pickering Triangle," a beautiful region within the Veil nebula that despite its name was discovered by Williamina. Or the most famous of all: the Horsehead Nebula, a dark nebula that is seen against the light of the luminous nebula that is behind it, with the recognizable appearance of a chess horse head. It is an icon of astrophotography, which Williamina located on photographic plate No. B2312 (taken on February 6, 1888 by Pickering's brother William, also an astronomer at Harvard) and described succinctly as a "semicircular notch 5 minutes in diameter, 30 minutes south of Zeta." When the *New General Catalogue of Nebulae and Clusters of Stars* (NGC) was compiled, the astronomer John Dreyer attributed the discovery of this nebula (and of the other 58) to the director of the Harvard Observatory; however, Pickering himself clarified the discovery in the *Annals of the Observatory* of 1890 and this omission was corrected in the second edition of the NGC with Williamina receiving full credit as the author of the discovery.

In the summer of 1898, a gathering of distinguished astronomers took place at Harvard. Presentations on various topics were given including one written by Williamina on variable stars and their hydrogen lines. It was read by Pickering, who added at the end of the presentation that nearly all of the 79 stars presented were discovered by Williamina.

She discovered in total more than 200 variable stars. In 1910, she reached the peak of her scientific career: her system of spectral classification allowed her to discover a new class of star. Unknown until then, these would later be called "white dwarfs." White dwarfs are very dense and hot objects composed of the embers that remain when a normal star has consumed its nuclear fuel; they are undeniably dead stars. They have a mass comparable to that of the Sun, but their size nevertheless is similar to that of Earth; they shine thanks to the residual heat still stored from when they "burned" with nuclear reactions (burning is, of course, a chemical reaction, while stars shine due to nuclear reactions). In addition to the many nebulae and stars Williamina discovered, we must also add ten novae (a nuclear explosion in a white dwarf star).

Williamina worked at the Harvard Observatory until her death in 1911, aged fifty four. Like Draper, she died of pneumonia; it was still seventeen years before another Fleming, Alexander, discovered penicillin. Her obituary was published in several scientific journals, including the prestigious *Science*, written by her colleague and friend, Annie J. Cannon: "Of a large-hearted, sympathetic nature, and keenly interested in all that pertains to life, she won friends easily, while her love of her home and unusual skill in needlework, prove that a life spent in the routine of science need not destroy the attractive human element of a woman's nature."

She received many honors in life for her work dedicated to science, and was made a member of the British Royal Astronomical Society (the sixth female member of this institution in its history), of the Astronomical and Astrophysical Society of America, and of the Société Astronomique of France. She received the "Guadalupe Almendaro" medal from the Astronomical Society of Mexico shortly before her death. In her time, she was the most famous astronomer in America. As for her son, Edward P. Fleming, he graduated from MIT in 1901 and worked as a mining engineer in Chile. A planetary nebula, "Fleming 1," containing two white dwarfs was named in honor of Williamina, and also (together with Alexander Fleming—though the two are unrelated) a large crater of 106 km in diameter on the Moon, located on the far side.

Figure 56 Location of crater Fleming. Courtesy of the Lunar and Planetary Institute, Houston, Texas.

Figure 57 Lunar Reconnaissance Orbiter zoom on crater Fleming (image width is 150 miles).

13

Annie Jump Cannon (1863–1941)

Figure 58 Annie Cannon (photo taken 1922).

Oh, Be A Fine Girl, Kiss Me!
Mnemonic for spectral
classification of stars,
a technique originated by Cannon

Annie Jump Cannon was born on December 11, 1863 (under the sign of Ophiuchus), in Dover, Delaware, the eldest daughter of Wilson Lee Cannon, a wealthy shipbuilder and state senator, and his second wife, Mary Elizabeth Jump. As a child, she observed the sky with her mother, who had taken an astronomy course, using a small telescope.

At age 16 she entered Wellesley College (founded in 1870 as the Wellesley Female Seminary), which by then offered an excellent program of studies in astronomy. Her father, always the politician, prepared the way by sending Wellesley's president a box of delicious peaches. In the course of her studies at Wellesley she studied under Mary Whitney, herself a student of Maria Mitchell at Vassar, and Sarah F. Whiting (1847–1927), a professor of physics who had visited Pickering at Harvard to learn about the new techniques of stellar spectroscopy. Annie graduated in 1884 and returned to Dover. A childhood case of scarlet fever caused her to become almost completely deaf, but she could read lips. There, for the next ten years she devoted herself to music, photography, and enjoying social life. She undertook a trip to Europe in 1892 and published a book of her photographs (a new art at the time) including the first photograph of Córdoba Mosque. Her life

The Women of the Moon. Daniel R. Altschuler Stern and Fernando J. Ballesteros Roselló.
© Daniel R. Altschuler Stern and Fernando J. Ballesteros Roselló 2019. Published in 2019
by Oxford University Press. DOI: 10.1093/oso/9780198844419.001.0001

changed with the death of her beloved mother in 1893. She decided to return to Wellesley to work with Whiting and a year later she enrolled at Radcliffe College (Harvard's affiliate for women) where she studied astronomy for two years with Professor Pickering.

In 1896, she joined Pickering's computer team, to work with Williamina Fleming and Antonia Maury (the subject of our next chapter) studying stellar spectra and cataloging variable stars.

At the foundation of experimental science, we find taxonomy— ordering what is observed in categories that in some way help us understand the nature of what is studied. The categories are initially arbitrary, and many do not lead to anything interesting. Sorting people according to the alphabetical order of the first letter of their last name does not lead us to understand human beings, however useful it may be for the telephone directory. On the other hand, classifying people according to characteristics of their mitochondrial DNA allows us to understand kinship relationships between different populations of humans. Similarly, classifying stars according to their apparent brightness in the sky does not inform us about their nature, since the distance at which a star is located is an arbitrary factor that influences its apparent brightness.

There were several star classification systems before Annie's work, but they were nothing more than a phone book. By classifying the stars according to the presence of certain spectral lines, Annie opened the door to a more fundamental understanding.

Cannon had an uncanny ability to classify the spectra of the different stars. Cecilia Payne-Gaposchkin (1900–1979), who started working at Harvard in 1927, said that Annie did not analyze spectra, she sat down with her glass plates in their wooden frames, and simply recognized the spectra as if she were recognizing faces; she could classify three stars in a minute. Annie managed to catalog by spectrum over 350,000 stars from all over the sky, a titanic task summarized in the Henry Draper Catalogue publication between 1918 and 1924, and annexes published later. Annie reordered and simplified the spectral classes of Fleming and Pickering, which had initially been designated by the letters A, B, C, etc. Finding that the categories correlated with the temperature of the star if she ordered them as follows: O, B, A, F, G, K, M (students use the mnemonic rule at the head of this chapter to remember this order) with subcategories indicated by a number.

Our Sun, for example, is a star of spectral type G2, in this still-used typology. Later, Payne-Gaposchkin established the theoretical basis of this classification system. The work of spectral classification, pioneered by Cannon and Payne-Gaposchkin, marked the beginning of modern astrophysics, including the then-novel idea that the stars were composed of hydrogen. The following table summarizes the classification system.

Class	Temperature (K)	Color	Mass[1]	Luminosity	Main absorption lines	Example
O	28,000–50,000	Blue	20–60	90,000–800,000	Nitrogen, carbon, helium, and oxygen	Naos
B	10,000–28,000	Blueish white	3–20	95–90,000	Helium and hydrogen	Rigel
A	7500–10,000	White	2–3	8–95	Hydrogen	Sirius A
F	6000–7500	Yellowish white	1.1–1.6	2–8	Metals: iron, titanium, calcium, strontium, and magnesium	Polaris A
G	4900–6000	Yellow	0.8–1.1	0.6–2	Calcium, helium, hydrogen, and metals	The Sun
K	3500 4900	Orange Yellow	0.6–0.8	0.1–0.6	Metals and titanium oxide	Alfa Centauri B
M	2000–3500	Red	0.w.8	0.001–0.1	Metals and titanium oxide	Gliese 581

[1] The mass, and luminosity, are given in proportion to that of the Sun (Sun = 1).

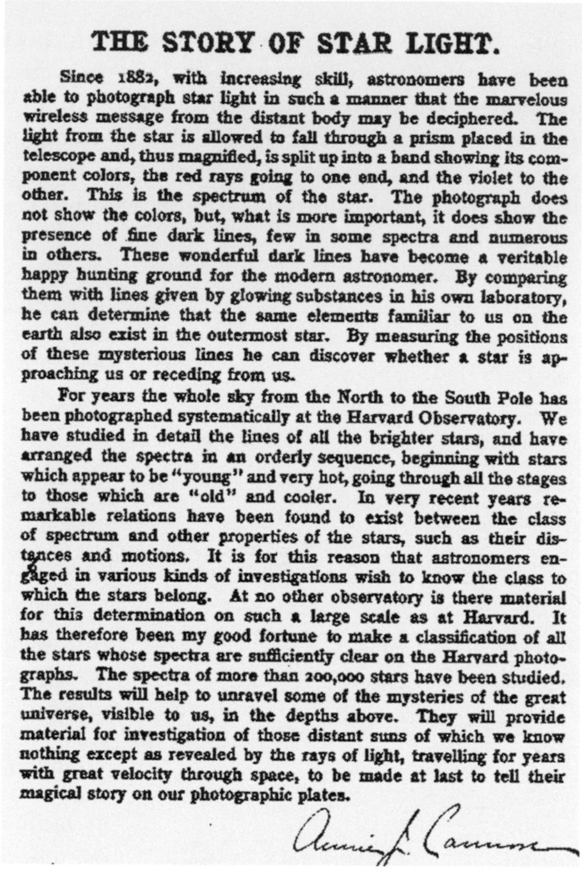

THE STORY OF STAR LIGHT.

Since 1882, with increasing skill, astronomers have been able to photograph star light in such a manner that the marvelous wireless message from the distant body may be deciphered. The light from the star is allowed to fall through a prism placed in the telescope and, thus magnified, is split up into a band showing its component colors, the red rays going to one end, and the violet to the other. This is the spectrum of the star. The photograph does not show the colors, but, what is more important, it does show the presence of fine dark lines, few in some spectra and numerous in others. These wonderful dark lines have become a veritable happy hunting ground for the modern astronomer. By comparing them with lines given by glowing substances in his own laboratory, he can determine that the same elements familiar to us on the earth also exist in the outermost star. By measuring the positions of these mysterious lines he can discover whether a star is approaching us or receding from us.

For years the whole sky from the North to the South Pole has been photographed systematically at the Harvard Observatory. We have studied in detail the lines of all the brighter stars, and have arranged the spectra in an orderly sequence, beginning with stars which appear to be "young" and very hot, going through all the stages to those which are "old" and cooler. In very recent years remarkable relations have been found to exist between the class of spectrum and other properties of the stars, such as their distances and motions. It is for this reason that astronomers engaged in various kinds of investigations wish to know the class to which the stars belong. At no other observatory is there material for this determination on such a large scale as at Harvard. It has therefore been my good fortune to make a classification of all the stars whose spectra are sufficiently clear on the Harvard photographs. The spectra of more than 200,000 stars have been studied. The results will help to unravel some of the mysteries of the great universe, visible to us, in the depths above. They will provide material for investigation of those distant suns of which we know nothing except as revealed by the rays of light, travelling for years with great velocity through space, to be made at last to tell their magical story on our photographic plates.

Annie J. Cannon

Figure 59 Christmas card "The Story of Starlight" by A. Cannon.

In 1915, Annie described her work on a Christmas card that she titled "The Story of Starlight," shown in Figure 59.

Annie received widespread recognition for her work. In 1914, she was appointed honorary member of the Royal Astronomical Society, and in 1921, she received her doctorate in astronomy from the University of Groningen, in the Netherlands. In 1925, she received an honorary doctorate from Oxford University, the first woman to do so. In 1931, she received the prestigious Draper Medal from the National Academy of Sciences of the United States. She shares this distinction with others such as A. Michelson, P. Zeeman, A. Eddington, and H. Bethe. In 1938, she was finally appointed to the William Cranch

Bond Chair in astronomy at Harvard (the title honors the first observatory director, and friend of Maria Mitchell's father). Her appointment letter was headed with "Dear Sir".

Annie worked at the observatory for forty-five years until her passing in 1941 at the age of seventy-seven. Her lunar crater (56 km diameter) was assigned in 1964 by the IAU. It is located to the northwest of Mare Marginis (Sea of the margin), near the northeast margin of the visible face of the Moon. Note the nearby crater Joliot, named for Jean Frédéric Joliot-Curie (excluding Irène Curie), and also, to the southwest of Joliot, Hubble, named for Edwin Hubble. For Cecilia there is no lunar crater, although she is recognized as the most illustrious of the Harvard women, her work being one of the most cited in the scientific literature.

Figure 60 Location of crater Cannon. Courtesy of the Lunar and Planetary Institute, Houston, Texas.

Figure 61 Lunar Reconnaissance Orbiter zoom on crater Cannon (image width is 150 miles).

14

Antonia Caetana de Paiva Pereira Maury (1866–1952)

Figure 62 Antonia Maury. Harvard College Observatory.

And they that scan the heavens by night,
Since truth's clear light they saw,
No human meets and measures serve,
But Nature's mightier law.

From ANTONIA MAURY's poem
dedicated to the Vassar College
Observatory in 1896[1]

Antonia Caetana de Paiva Pereira Maury[2] was another of the Harvard Observatory's computers, dedicated to the analysis and classification of spectra for the Henry Draper Catalogue. But she was not just another computer of the team, since her uncle was the same Henry Draper, who gave the name to the catalog.

She was born in New York in 1866 to a distinguished family, a year after the end of the American Civil War. Despite the difficulties of the postwar period, she was immersed in a scientific environment all her life. Her maternal grandfather was John William Draper (1811–1882), chemist, historian, and amateur astronomer, and author of the first photographs of the Moon—daguerreotypes taken through a telescope, although they have not been preserved. His son was the aforementioned Henry Draper, doctor, chemist, and amateur astronomer at the

---

[1] Antonia C. Maury (1923). Verses to the Vassar Dome (a poem). *Popular Astronomy*, Vol. 31, p. 76.

[2] Her mellifluous name was bestowed in honor of her Portuguese paternal grandmother, Antonia Caetana de Paiva Pereira, daughter of the physician to the Emperor John VI of Portugal.

The Women of the Moon. Daniel R. Altschuler Stern and Fernando J. Ballesteros Roselló.
© Daniel R. Altschuler Stern and Fernando J. Ballesteros Roselló 2019. Published in 2019 by Oxford University Press. DOI: 10.1093/oso/9780198844419.001.0001

University of New York. Antonia spent much of her childhood in the laboratory with her uncle, helping him with his test tubes, and probably this experience and her grandfather infected Antonia with her love of astronomy. Her father, too, added to the scientific ambience: the Reverend and Professor Mytton Maury was a well-known naturalist. He instructed her at home with her sister Carlotta, and taught them not only to read and write (at nine years of age Antonia already read Latin with ease), but also mathematics and sciences, including field trips to collect specimens, observation and recognition of the night sky, and the principles of photography, which was a leading technology at that time. In short, Mytton taught them everything they needed to be well educated and independent.

This propitious family environment led Carlotta to become a paleontologist, one of the first women working in this discipline. Her brother, William, became a surgeon while her cousin Matthew Fontaine Maury (older than them), commander of the United States Navy, was a renowned oceanographer (nicknamed "the explorer of the seas") and astronomer. With such a background it is no surprise that he became the first director of the United States Naval Observatory. And, of course, it nurtured Antonia herself.

She entered Vassar College in 1883, Yale University's sister institution (Yale was then only open to male students)[3]. Vassar was and remains one of the most selective and prestigious in the United States. Given the scientific preparation of the young Maury, she excelled, especially in philosophy and astronomy, and soon became one of the school's most distinguished students. Her time there was a happy one; she made observations from the dome of the Vassar College Observatory, led by Professor Maria Mitchell (whom we have already met) and enjoyed an atmosphere of intellectual freedom that later she would miss[4]. She graduated in 1887 with honors in physics, astronomy, and philosophy. Being especially well suited for astronomy, there was a natural place for her to go: the Harvard Observatory, which had begun the year before to work on the Henry Draper Catalogue. In 1889, therefore, she

[3] The first woman was admitted to Yale in 1892. The first male student was admitted to Vassar in 1970.

[4] Almost a decade later, in 1896, when she attended the traditional festival of the dome of the department of astronomy, she read a long poem she had composed dedicated to the Vassar College Observatory (a fragment of which is shown at the beginning of this chapter) that showed the happiness of that time.

went to work with Pickering's group of astronomers, happy to continue the business that her uncle had started. But life there was not easy.

Antonia's relentless focus on intellectual matters manifested itself in a carelessness about her physical appearance and dress (like Noether, as we will see): she showed holes in the heels of her stockings and did not worry about her hairstyle. This whole package was too much for her uncle's wife, Mary Anne Draper, who financed the work of the Pickering group and was an elegant woman of high society. In addition, Antonia was very independent and did not like to work with the other women in the observatory. This provoked taunts and qualms, and since she was Draper's niece, many saw her as "plugged in"—despite the very bad impression her aunt had of her! She made few friends at the Harvard Observatory, although the ones she made appreciated her highly; according to Dorrit Hoffleit, "she was a lady cultivated in everything except her personal appearance" who "could talk about any imaginable subject."

Cecilia Payne-Gaposchkin recalled Maury in her autobiography[5]:

> Like Miss Cannon, Miss Maury was extremely good to me. Many were the long talks that we had about the problems of stellar spectra. We both liked to work at night, and our discussions were painfully punctuated by insect bites, for she insisted on keeping the windows open and could not bear to kill the mosquitoes. I have often speculated on the impetus that is given to a person by some handicap of nature. Miss Cannon and Miss Leavitt were both deaf. Miss Maury's handicap was extreme homeliness, mitigated by her beautiful brown eyes. She did not, indeed, pay much attention to her appearance. She would come to work, her associates said, in one black stocking and one brown. In this she was a contrast to Miss Cannon, who was always smartly dressed, and even in old age was very handsome. Her pictures show that in youth she was extremely beautiful.

Antonia Maury was especially gifted for theoretical work and was one of the most intellectually capable persons on the team. She did not want to be a passive observer, she wanted to *understand* what she was observing—she had a passion for understanding things. She argued about science with Pickering often; one of their first disputes arose following her first important discovery during the elaboration of the Henry Draper Catalogue: the spectral binaries.

[5] Cecilia Payne-Gaposchkin (1996). *An autobiography and other recollections*. Cambridge University Press.

A "binary" is a pair of stars that are so close together that, from Earth, you cannot see them separately without proper instrumentation. Sometimes a telescope suffices to distinguish the two stars, but not always. How to know, then, if there really are two stars close together or only one? Looking at the spectrum, was the recently discovered solution in the 1880s. In 1887, Pickering discovered that the spectrum of the star zeta Ursae Majoris actually looked like the mix of two different spectra. But only a single point of light was visible through the telescope. He wondered if they were not actually two different stars, so close together that they could not be distinguished. It turned out that this was the case: the first spectral binary had been discovered.

The first job that Pickering assigned to the recently arrived Antonia Maury was to calculate how long it took for these two stars to complete an orbit, one around the other. During this investigation, Antonia found another star, beta Aurigae, whose spectrum was also the mixture of two different spectra: she had found the second spectral binary in history. The two discoveries were published in an article authored by Pickering in the *American Journal of Science* (1890), with a note at the beginning stating that "a detailed study of the results has been made by Miss A.C. Maury a niece of Dr. Draper", but without mentioning her in the rest of the article, nor detailing her (essential) contributions to the research, which provoked Antonia's anger. The discovery of beta Aurigae itself appears in a short addendum without attribution[6].

After this episode, Pickering assigned her to the classification of the spectra of the bright stars of the northern celestial hemisphere. She had to look at all the photographic plates of these stars and catalog them using the observatory's classification system, initially designed by himself and Williamina Fleming, and later simplified by their collaborator Annie Cannon. This was a huge and monotonous job that was not well suited for the inquisitive mind of Maury. But during this work, Antonia made a discovery: for some stars the spectral lines were narrow and for others, wide.

Pickering did not give the least importance to the discovery and attributed those differences to different conditions of observation at the time the spectra were taken: perhaps the ambient temperature or humidity affected the photographic plate, or perhaps a little cloud passed in front of the telescope, but Antonia thought it was a characteristic

---

[6] Edward C. Pickering (1890). On the spectrum of zeta Ursae Majoris. *American Journal of Science*, Vol. 39, p. 46.

of the star. And since there was no way to include this information in Annie Cannon's classification system, she discarded it completely and created her own. She defined twenty-two groups, depending first on the temperature of the star, and adding a label: *a* for spectra with well-defined broad lines; *b* for spectra with blurred broad lines; and *c* for spectra with narrow well-defined lines.

This led to a new series of confrontations with Pickering, who did not approve of this new classification system, and considered it a source of delay in the collective work which undermined its main objective: obtain quick results and finish the Henry Draper Catalogue as soon as possible. He did not even encourage women to do original work, although he allowed it so long as it did not hinder the progress of the project, and at this point, in his opinion, Antonia was imperiling the schedule for the larger project.

The friction between the two was frequent. As she wrote in her diary, she complained that she always wanted to "study calculus, but Professor Pickering did not want it." She did not feel supported, and he did not welcome her attempts to understand the observations. Her aunt was, it is fair to say, the opposite of supportive, writing to Pickering that "she is not a valuable member of the team. I'll be happy when you get rid of that nuisance." But it was not necessary to reach that extreme. It was Antonia herself who, only two years after entering work for Pickering, could not take the environment any longer. She decided to leave the observatory in 1891 to devote herself to teaching, lecturing at Cornell University and other universities in New York and also at the Gilman School in Cambridge, Massachusetts, and later at Miss Mason's school in Tarrytown, New York, an activity to which she would dedicate herself for more than twenty years.

But as she had left her draft spectral classification incomplete, she returned sporadically to the observatory to finish it. Her work was finally completed with the publication in 1897 of her catalog in the *Annals of the Observatory*, an encyclopedic solo work of 128 pages in which she focused on the mystery of the width of the spectral lines that fascinated her so much.

After its publication, and although her main occupation was teaching, she continued to dedicate herself in her free time to astronomy, studying other spectroscopic binaries (in particular, the beta Lyrae system, which was a mystery and on which she worked for several years, studying almost 300 spectra of this system). But nothing seemed to indicate that her work at the Harvard Observatory would be of any importance.

It would thus remain fairly obscure, a niche observation without much consequence, until in 1905, a Danish astronomer, Ejnar Hertzsprung (1873–1916), discovered that there were stars that, although they had the same appearance to the telescope, were much brighter than others[7]. The only possible explanation was that they were stars with the same temperature but different size; the brightest must have been enormously large, and he called them "giants." When he stumbled upon the work of Antonia Maury, he discovered that the stars she had tagged with the characteristic c, that is, very narrow lines in the spectrum, coincided with the stars that he had identified as giant stars. What her discovery measured, in short, was the size of the stars.

Ejnar Hertzsprung wrote that, in his opinion, Antonia Maury's classification was "the most important advancement in stellar classification since the trials by Vogel and Secchi . . . To neglect the c-properties in classifying stellar spectra, I think, is nearly the same thing as if a zoologist, who has detected the deciding differences between a whale and a fish, would continue classifying them together." Hertzsprung wrote to Pickering and urged him to include Antonia's classification scheme in Draper's official catalog, which he however did not do. As Dorrit Hoffleit put it, Pickering's issue "was not that he could not see it, it was that he was upset because it was she who had discovered it".

But the International Astronomical Union did, and in 1922 modified its classification scheme (based on that of Annie J. Cannon), adding the prefix c for stars that had narrow lines in their spectra. And much later, in 1943, the astronomers Morgan, Keenan, and Kellman, who developed the "MKK system" in the revised Henry Draper Catalogue, fully adopted Maury's classification system.

Antonia returned to Harvard as an adjunct professor in 1918, and the next year Pickering died (although one thing does not have anything to do with the other). But it is true that she had better relations with the new director of the observatory, Harlow Shapley, and continued her research in binaries. She discovered another phenomenon that could also

---

[7] This discovery was also made independently by another astronomer, the American Henry Norris Russell (1877–1957). The graph plotting out this phenomenon, one of the most important in the history of science, is called the Hertzsprung–Russell diagram

alter the width of the lines of a star's spectrum: rotation. She saw that a fast rotation was correlated with wider spectral lines. During the following years, she practiced her vocation with pleasure, and worked at the observatory until her retirement in 1935, at almost seventy years of age.

After retiring, and emulating her father, she dedicated herself to ornithology, botany, and environmental conservation, and saved with her activism the redwood forests that, during the Second World War, the United States government wanted to cut down for wood needed for its army. She was furthermore placed in charge of the museum of Draper Park's observatory complex built by her uncle and her grandfather decades before. She was a member of several scientific societies, such as the American Astronomical Society, the Royal Astronomical Society, and the National Audubon Society (ornithology). And at the age of seventy-seven she received an award full of irony: the Annie J. Cannon Prize for her stellar classification system, an award that her former colleague (precisely the author of the classification system that Antonia discarded) had instituted in 1933 for those women residing in the United States who had made an important contribution in the field of astronomy (Cecilia Payne-Gaposchkin was the first recipient).

Independent until the end, she never married. She died peacefully in 1952, at eighty-six. In her honor and that of her cousin, Commander Matthew Fontaine Maury, the International Astronomical Union named Maury: a small, well-defined crater almost 18 km in diameter next to the Lagoon of Dreams (Lacus Somniorum), to the northeast near the edge of the visible disk.

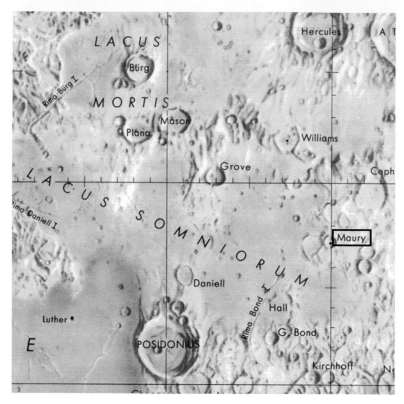

Figure 63 Location of crater Maury. Courtesy of the Lunar and Planetary Institute, Houston, Texas.

Figure 64 Lunar Reconnaissance Orbiter zoom on crater Maury (image width is 150 miles).

15

Henrietta Swan Leavitt
(1868–1921)

We shall never understand it until we find a way to send up a net and fetch the thing down!

Exclamation of HENRIETTA LEAVITT during her work at the Harvard Observatory, referring, exasperated, to the strange behavior of beta Lyrae[1]

Figure 65 Henrietta Leavitt (photo of 1921).

Henrietta Swan Leavitt[2] was born in 1868 in Lancaster, Massachusetts, into a Puritan family—one of the first colonial New England families, in fact—of deep religious convictions, convictions that she would keep for all her life. We know little about her childhood, except that she had six siblings (only two would reach adulthood) and that, due to her father's profession as a pastor, she spent much of her childhood traveling to the various parishes to which he was assigned, at various places in Cambridge, and later in Cleveland.

Her life is largely a blank until 1888, when she enrolled at Radcliffe College at twenty years of age, somewhat older than usual, perhaps due to the itinerant life of her father. It does not seem that Henrietta had initially any inclinations towards science, but that she acquired her taste for astronomy by chance when in 1892, in her last year at the university, she decided to enroll in an astronomy course. This subject fascinated

---

[1] George Johnson (2005). *Miss Leavitt's stars*, p. 88. Norton.

[2] George Johnson (2005). *Miss Leavitt's stars*. Norton.

The Women of the Moon. Daniel R. Altschuler Stern and Fernando J. Ballesteros Roselló.
© Daniel R. Altschuler Stern and Fernando J. Ballesteros Roselló 2019. Published in 2019 by Oxford University Press. DOI: 10.1093/oso/9780198844419.001.0001

her in a way that others had not, and once she graduated, she decided to dig deeper into the subject and enroll in a two-year postgraduate course in astronomy at the Radcliffe College Observatory.

Those two years of graduate school finally brought out her vocation. After finishing her course of studies and knowing that Pickering at the Harvard Observatory was recruiting women to carry out the most ambitious stellar classification project in the world, she decided to offer her services as a volunteer. Since free labor was always welcome, Henrietta began work at the Harvard Observatory as an unpaid assistant in 1895.

Henrietta's health was delicate, and she was afflicted with a serious illness, probably meningitis, that sent her home shortly after entering the observatory. As a result, she was deaf for the rest of her life. Her convalescence was long and she spent several years at home. During this time, Pickering did not forget her and continued to exchange correspondence with her, asking about her health and keeping her well informed about progress at the observatory. By 1902, Henrietta was quite recovered, and wrote to the director of the observatory about the possibility of returning to her volunteer position, but working from home. Pickering wanted to have all his team together at the observatory, because this facilitated the progress of the stellar classification project. He therefore persuaded her to return to the observatory in exchange for offering her a permanent paid position. Henrietta accepted and moved to live with her uncle Erasmus at a village near the observatory.

At the observatory she made friends with the other women computers. Her companions remember her as a person with a sweet, hard-working character and an optimistic temperament, "who saw beauty everywhere." A religious and devout person, her puritanical education led her to be completely dedicated to her work, not shunning hard work where others could profit off her results. As with the other computers, Pickering gave Henrietta few opportunities to use her talent in theoretical studies. But she (unlike Antonia Maury) never complained. She did not protest, therefore, when he assigned her the task of performing the photometric study of the catalog plates, that is, measuring how much light really came from each star, a branch of astronomy that was less "glamorous" than the newest spectroscopy to which the other ladies were dedicated. It was an arduous and thankless task.

The response of the photographic plates with which astronomical images were made at that time was not proportional to the brightness of the stars. A star that was twice as bright as another did not leave an image twice as intense on the photographic plate, but rather less. In addition, the photographic impression varied with the star's color: a bluish star and a reddish star that were of equal brightness would not leave the same imprint on the photographic plate, since the emulsions of silver salts were more sensitive to the blue color. How to deal with this problem? The idea was to measure as precisely as possible how much light came from a set of stars, seeing how the different plates registered these stars, and then calculating the light for the rest of the stars in the sky by a simple comparison with this set. In astronomy these are called standard stars.

Due to the dedication of Henrietta to her work, Pickering assigned her to perform these calculations with a set of stars near the north celestial pole (and therefore, visible any night of the year), which would become the standard stars to use with the rest of the photographic plates. This set was called the *North Polar Sequence*. To do this she had to compare the images taken with thirteen different telescopes! Cecilia Payne-Gaposchkin, Henrietta's partner at the observatory, said[3]: "It may have been a wise decision to assign the problems of photographic photometry to Miss Leavitt, the ablest of the many women who have played their part in the work of Harvard College Observatory. But it was also a harsh decision, which condemned a brilliant scientist to uncongenial work".

In 1911, at what would prove to be nearly the midpoint of this gargantuan effort, Leavitt's father died, and the main source of income for the family was lost. Henrietta brought her mother to live with her and the two moved to a small house in Cambridge that she could barely afford with her salary from the observatory. Economic hardship and health problems would plague her for the rest of her life.

The laborious process of measuring stellar brightness took fifteen years, from 1902 to 1917, the year in which the results were finally published in the *Annals of the Observatory*, and became her most famous publication during her life. The work was important for cataloging the relative brightness of stars, a useful but pedestrian reference. History, however, had its eyes on another aspect of her research.

Given her great ability measuring starlight, she devoted herself to studying variable stars, those whose brightness is not constant but changes over time, and she discovered a huge number of these stars

---

[3] Katherine Haramundanis (1996). *Cecilia Payne-Gaposchkin: an autobiography and other recollections*, p. 146. Cambridge University Press.

(about 2400). She focused her attention on the only two properties she could identify about these stars: their brightness and their period of variation, and she asked herself: could there be any relationship between these characteristics?

The crux of the problem was that she could not know what the real brightness of a star was; she could only measure the *apparent* brightness, what it seems to be when we see the star from Earth. A very bright star may give tenuous light because it is far away, and a dim one may seem very bright because it is very close. How can we find out its real brightness (the absolute luminosity) if we do not know how far away it is? Measuring the distance to celestial objects is one of the most important and difficult problems of astronomy, even today. At that time, you could only measure the distances to the nearest stars with anything like precision by calculating how they changed their apparent positions in the sky when the Earth moved from one side to the opposite side of its orbit (measuring the parallax).

The only way, Henrietta reasoned, for us to compare the brightness of different stars is to know that the stars in question are together in the same place in the universe. Thus, although we may not know how far they are, since they are at the same distance from us, we can compare their brightness.

Studying a series of images of the Small Magellanic Cloud[4] taken from the Harvard station at Arequipa, Peru, she found twenty-five stars that varied in a very similar way to a much closer star, the variable star delta of the constellation of Cepheus. These stars generically receive the name of Cepheid variables. Cepheids vary in brightness in a period ranging from one day to three months. We know today that these changes are due to pulsations of the star, which as it depletes the nuclear fuel at its center, becomes unstable, and evolves into the red giant phase.

Henrietta had found twenty-five Cepheids in the Small Magellanic Cloud, close enough to each other (in relation to their distance to the Earth) that she could compare them. What she discovered in this com-

---

[4] Today we know that the Small Magellanic Cloud (approximately 200,000 light-years away) and the Large Magellanic Cloud (approximately 160,000 light-years away) (named after the navigator Ferdinand Magellan who saw them in the southern hemisphere's sky during his circumnavigation of the globe) are two small satellite galaxies of the Milky Way. Despite this relationship to our Galaxy, the Small Magellanic Cloud is so far away that, by comparison, all the stars in it can be considered to be at the same distance from us.

parison was that they exhibited a strange property: the brighter the star was, the longer the oscillation period of its luminosity. In fact, stars' brightness was directly proportional to the period of variation: if one Cepheid star shone twice as brightly as another, its light also oscillated with a period that was twice as long as the other. It was a spectacular discovery, since it could provide a method to directly measure the distance to the farthest stars. This revolutionary work was published in 1912, but Henrietta's natural prudence led her to title her work modestly: "Period of the 25 variable stars of the Small Magellanic Cloud"[5]. In the text of the article was the important statement: "A remarkable relation between the brightness of these variables and the length of their periods will be noticed [. . .] Since the variables are probably at nearly the same distance from the Earth, their periods are apparently associated with their actual emission of light, as determined by their mass, density and surface brightness."

Although the report was signed by Pickering, as was customary, the first sentence states "The following statement regarding the periods of 25 variable stars in the Small Magellanic Cloud has been prepared by Miss Leavitt."

That same year, 1912, saw Leavitt falling ill once again, and she spent a year convalescing. Her work attracted the attention of the American astronomer Harlow Shapley (1885–1972), who wrote to her during that period and asked her if she could check if the same phenomenon

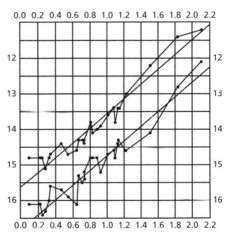

Figure 66 The original period–luminosity relation (horizontal axis logarithm of period in days/vertical axis more luminous smaller numbers) as published by Leavitt. The two lines correspond to the maximum and minimum brightness of each star.

[5] Edward C. Pickering (1912). Periods of 25 variable stars in the Small Magellanic Cloud. *Harvard College Observatory Circular*, Vol. 173, pp. 1–3. Pickering acknowledges at the beginning that the work is by Miss Leavitt.

happened in the Large Magellanic Cloud; in short, treating her as a colleague. She could not, unfortunately, answer—her ailments prevented her from doing the job that he asked for. Nor could she dig deeper into the subject of Cepheid variables. As she left it, her discovery only allowed one to know if one Cepheid was closer than another. It was the task of others to perform the proper calibration so that Cepheids could be used to measure absolute distances. In particular, the Danish astronomer Ejnar Hertzsprung (1873–1967) and Shapley himself (who would become director of the Harvard Observatory) were dedicated to accurately measuring the parallax of a nearby Cepheid and determining its distance. Once the distance to that Cepheid was well established, it would then be possible to know how far away the others were by a simple rule of proportionality.

One of the main scientific debates of that time was over the question of the actual size and shape of the universe. There were two opposing positions: one defended the view that the whole set of stars that we see in the sky and that seem to be grouped in a large disk (what we call the Milky Way) was the whole universe; and another that believed that the weak spiral-like nebulae that could be seen through telescopes (some of which showed stars inside) were actually other structures similar to our Galaxy but very distant, other "island-universes", as they were called at the time, using Immanuel Kant's term.

Shapley initially defended this second position. But in 1918, in a crucial study, he used the tool recently created by Henrietta Leavitt to measure the size of the Galaxy. Locating the most distant Cepheids he could find in the Milky Way, he calculated that they were at an incredibly large distance: the Milky Way at hundreds of thousands of light-years across, was much larger than anyone dared to imagine. Therefore, it must be the whole universe, and should include the spiral nebulae.

This scientific question materialized in "The Great Debate", which took place in April 1920, at the headquarters of the United States National Academy of Sciences. The representatives for each position were precisely Harlow Shapley, who, as a result of his measurements defended the idea that the Milky Way was the whole universe (and that the spiral nebulae were small structures within the Milky Way), and the astronomer Herbert D. Curtis, who defended a size for the Milky Way ten times smaller and argued that the spiral nebulae were other galaxies analogous to ours.

There were earthly as well as celestial matters to be settled at the debate: the year before, Pickering had died, and the Harvard administration had come to the debate because they were interested in offering Shapley the directorship of Harvard. He was keenly aware of this and perhaps for that reason he did not want to be very aggressive or radical in his reasoning. With Shapley staying on well-established terrain—and probably giving Curtis a tactical advantage—the debate ended with a nominal triumph for Curtis. Despite this "loss," Harvard ended up offering the position to Shapley.

Upon assuming the direction of the Observatory, in 1921, the first thing he did was to promote Henrietta, naming her head of the department of photographic photometry. Unfortunately, she would enjoy this position only for a short time. That same year she fell ill again, although this time the disease would be serious: she was diagnosed with stomach cancer. She died on a rainy day in December of that same year, visited by Shapley and with Annie Cannon by her side. After a life filled with ailments, earning the salary of an unskilled worker for almost thirty years, at her death she left her mother her poor inheritance valued at $314. In life she received no honors or public recognition.

But that was not the end of her story. Curtis had "won" the great debate, but had he been right? Actually, it was Shapley who had correctly estimated the true size of the Milky Way[6]. But very soon it was going to be proven that in another aspect Curtis was right: the spiral nebulae *were* other galaxies. The universe, it would be shown, was almost unimaginably huge—even for astronomers accustomed to thinking in light-years!

The discovery of the size of the universe would lean heavily on Leavitt's work, but it would fall to Edwin Hubble (1889–1953), another American astronomer, just entering the fray. In 1923, from the Mount Wilson Observatory (for three decades the largest telescope in the world), he studied the largest of those spiral nebulae, the one located in the constellation Andromeda. Inside it, he discovered a Cepheid variable. Surprisingly, Hubble saw that the Cepheid variable he had found in the Andromeda nebula showed a very weak apparent brightness; using Leavitt's technique, he calculated a distance of 800,000 light-years for the nebula, placing it well outside the Galaxy. The nebula had to be a whole independent galaxy in itself! Hubble's later work showed that *everything*

---

[6] In addition, Shapley defended the position that the Sun was in a peripheral zone of the Galaxy, while Curtis believed that it was very close to its center. Shapley had also been right on this point.

that was previously called "spiral nebulae" turned out to be galaxies analogous to ours, but incredibly distant. The universe revealed itself to be much larger than previously thought[7]. Hubble wrote to Shapley about this discovery in a letter dated February 19, 1924, who after reading it said to Miss Payne, who happened to be in his office: "Here is a letter that has destroyed my universe"[8].

In the following years, Hubble continued to measure, by means of the Cepheid variable technique, the distance to several galaxies. Upon a suggestion from Shapley, he also began to study their spectra and compare them with the calculated distances. In this way, he made a discovery that—even more than the size of the universe—completely changed our understanding of the cosmos: the light coming from far galaxies is reddened. And beyond that, the effect is larger the further away a galaxy is, in a phenomenon astronomers refer to as "redshift". And just as the whistle of a train sounds lower when the train moves away from us than when it approaches (the Doppler effect), redshift appears because the galaxies are moving away from us, and those farther away are moving faster[9]. In reality, all galaxies are receding from all the others. Hubble discovered that the whole universe is expanding, and that, therefore, in the past all the matter in the universe must have been within an incredibly small volume. He had discovered the Big Bang.

The tool that Henrietta Leavitt discovered allowed us to measure the universe, and ultimately to realize not only how inconceivably large it is, but that it had a beginning. Today she is the most famous member of "Pickering's harem", and although she was not well known during her lifetime, she became famous after her death. Even the Nobel Prize grazed her slightly: in 1925, Gösta Mittag-Leffler wrote to Leavitt: "I feel seriously inclined to nominate you to the Nobel Prize for 1926". Alas, it was five years since she had passed away and the Nobel is not awarded posthumously. Shapley answered the letter, informing him of her death and expressing respect for her work:

---

[7] Today the distance to the Andromeda nebula has been corrected to 2,200,000 light-years. The error comes from the fact, discovered later, that there are two different types of Cepheid variables.

[8] Dava Sobel (2016). *The glass universe*, p. 205. Viking.

[9] In a point almost too fine to quibble about, it is space itself that is expanding, carrying the galaxies away, but this does not change the story.

Miss Leavitt's work on the variable stars in the Magellanic Clouds, which led to the discovery of the relation between period and apparent magnitude, has afforded us a very powerful tool in measuring great stellar distances. [. . .] To me personally [the discovery] has also been of highest service, for it was my privilege to interpret the observation by Miss Leavitt, place it on a basis of absolute brightness, and extending it to the variables of the globular clusters, use it in my measures of the Milky Way. Recently, Hubble used the period-luminosity curve, which I obtained from Leavitt's work, to obtain the measurements of the distances of the spiral nebulae.

It is clear who Shapley thought deserved the award. Leavitt did not get the Nobel Prize or recognition in life, but she has slowly become known to the history of science. Gael Mariani[10] writes:

Little is known of Henrietta Leavitt's personal feelings about the way she had been overstepped. Hers was a shy and somewhat unassuming personality, and women at that time, even highly educated and brilliantly talented women who in a fairer world would have been respected as equals by their male peers, were all too often resigned to taking a lesser role, and were often just quietly grateful to be given any sort of role at all.

As has been wistfully observed, her deafness prevented her from hearing that a woman could not devote herself to astronomy. Her name was given to an asteroid, and to a crater 66 km in diameter on the far side of the Moon, next to the great Apollo crater.

---

[10] Gael Mariani (2012). Henrietta Leavitt—Celebrating the Forgotten Astronomer. AAVSO. https://www.aavso.org/henrietta-leavitt-%E2%80%93-celebrating-forgotten-astronomer

Figure 67 Location of crater Leavitt. Courtesy of the Lunar and Planetary Institute, Houston, Texas.

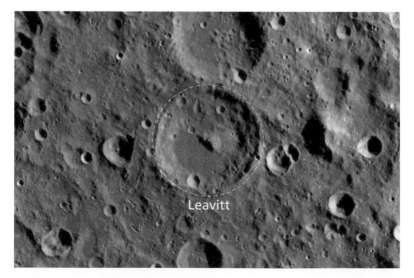

Figure 68 Lunar Reconnaissance Orbiter zoom on crater Leavitt (image width is 150 miles).

16

Mary Adela Blagg
(1858–1944)

Figure 69   Mary Adela Blagg. Cheadle and District Historical Society—photo *c.*1890.

Mary Adela Blagg was an extremely reserved woman, to the point that, despite having lived half of her life in the twentieth century, we only have one photo of her as a young woman. However, if anyone deserves to have her own crater on the Moon, it is certainly her, since she was the person who put order to the chaos of lunar nomenclature at the beginning of the twentieth century and laid the foundations of the nomenclature system used today by the IAU.

Blagg was born in 1858 in the town of Cheadle in the United Kingdom, where her father, Charles Blagg, worked as a solicitor and where she spent her entire life. Mary Adela received a basic education from her parents at home and then attended a private boarding school for young ladies in the neighborhood of Kensington, in London, where she studied to a level equivalent to secondary education. Despite her lack of formal training, she liked astronomy and mathematics[1], and following the spirit of Mary Somerville, she trained herself in this latter discipline thanks to the books she borrowed from her brothers. She not only nurtured her love for the subject, she also learned harmonic analysis, which would later prove very

---

[1]   And a sister discipline of the latter, chess, in which she became an expert.

The Women of the Moon. Daniel R. Altschuler Stern and Fernando J. Ballesteros Roselló.
© Daniel R. Altschuler Stern and Fernando J. Ballesteros Roselló 2019. Published in 2019
by Oxford University Press. DOI: 10.1093/oso/9780198844419.001.0001

useful. As for astronomy, although it always interested her, during her youth she never actively pursued nighttime observation.

She spent the first half of her life living in her parents' house without any trade or occupation beyond the domestic chores that she shared with her mother, reading about what interested her, performing her community volunteer work mainly in the environment of her parish, and the catechism classes she gave at Sunday school to earn some money. Nothing seemed to indicate that she was going to enter posterity with a crater on the Moon named after her.

The turning point came in 1903. At the age of forty-five, Mary Adela enrolled in a university extension course for adults on astronomy offered by Oxford University in her town, Cheadle. The speaker was the astronomer Joseph Alfred Hardcastle (1868–1917), grandson of John Herschel (the nephew of Caroline Herschel). He must have been an inspiring speaker, because Mary Adela suddenly discovered what she wanted to do with the rest of her life, although her lack of training was a serious drawback. During the course of the lectures, she approached Hardcastle and he suggested a topic where someone without scientific training could effectively collaborate and on which he had just started working: the Saunder project of lunar cartography.

Samuel A. Saunder (1852–1912), an amateur astronomer but widely known in professional circles, and a member of the Royal Astronomical Society (and whom we have already encountered in Chapter 2) had a private observatory in the town of Crowthorne on the outskirts of London, from which he had just embarked on a major project to expand and correct the lunar cartography[2] using higher-power telescopes. Hardcastle, after working for several years at the Oxford Observatory, had just moved to Crowthorne at the invitation of Saunder to collaborate with him on this project.

Similar to Harvard Observatory's large spectroscopy project, Saunder's lunar project also needed patient and intelligent people to do the laborious and painstaking but routine work that did not require special training. Through Hardcastle, Mary Adela got to Saunder and began her collaboration on the project, at first with some hesitation, but eventually becoming great friends with Saunder.

---

[2] Saunder had noticed that in the then-current maps, several craters were not correctly located, with errors of up to five seconds of arc, introduced in part by the libration movement of the Moon. Saunder tried to reduce the errors to the tenth of a second of arc, rendering the map fifty times more precise.

In the course of the lunar mapping project, they encountered reams of the two basic problems of the different systems of lunar nomenclature: in many cases the same structure had different names in each system, in others the same name was given to different structures. For that reason, in 1905 Saunder warned the Royal Astronomical Society about the problems caused by using different lunar nomenclatures. As we have seen, after a debate within the Royal Astronomical Society, it was agreed to carry out a comparative study of the three main nomenclatures—those of Mädler, Schmidt, and Neison—in order to ratify the agreements and resolve the contradictions. But who would do the work? Saunder did not hesitate to propose his friend and collaborator Mary Adela to make a comparative analysis of the three nomenclature systems. The proposal was accepted and an international commission was created, of which Saunder and Blagg were members, chaired by the director of the Oxford Observatory, Herbert Hall Turner (1861–1930).

Mary Adela began her comparative study of the works of Mädler, Schmidt, and Neison in 1907, a slow and cumbersome work that would take several years. Simultaneously, that same year, a series of unpublished observation books on variable stars that another astronomer, Joseph Baxendell (1815–1887), had made before his death thirty years prior fell into the hands of Turner. Turner decided to analyze these data to present them in the form of a publication. It was also another cumbersome job that was especially difficult because it was not always easy to identify the stars Baxendell had used as points of comparison. Turner needed help in the analysis of the data, and he wrote an advertisement requesting "skilled volunteers [. . .] to prepare this mass of valuable data for publication as soon as possible." The announcement was answered by Mary Adela Blagg, starting another fruitful collaboration with Turner, working simultaneously on the variable star booklets and on the problems of lunar nomenclature. This collaboration with Turner would result in the publication of ten joint articles on variable stars in the period 1912–1918. In the last article in the series, Turner would write that "practically all the editing work has been done by Miss Blagg. The difficulties of identification have appeared frequently and would not have been resolved without her care and patience." This work aroused Blagg's interest in variable stars, and with the experience gained she decided to tackle new research on such stars on her own, and study their periodicity, making use of the harmonic analysis she learned in her brothers' books. She published on her own six other articles with the results of these investigations[3].

As for the work on lunar nomenclature, in 1912, when it was already in the sixth year, tragedy struck: Saunder died. Just a year before he had retired and was working on the installation of a new private observatory near Oxford. He had just signed an agreement to rent the land in July when he fell ill with some illness that afflicted him throughout the rest of that year and that no doctor could diagnose, perhaps a cancer. His strength dwindled and he died in December of that year, four days before the meeting of the Royal Astronomical Society in which he was to be appointed president[4].

Blagg felt obliged to finish the task as soon as possible, in memory of her friend; even so, it still took her another year. In 1913, the *Collated List of Lunar Formations* was finally published, with Blagg and Saunder (as director) the credited authors. The work involved some 40,000 comparisons between the three catalogs and laid the foundations of the current selenography.

The following year the First World War broke out, and, like so many others, Mary Adela's scientific work was interrupted. During this period, she returned to her volunteer work, and devoted herself to the care of the Belgian children who had been welcomed in England as refugees of war. Nevertheless, in January 1916 she was elected to full membership of the Royal Astronomical Society with full rights (not honorary membership as had been extended to other women before—Somerville, Herschel, and Clerke), along with Annie Maunder, the first women to be elected, after the membership of women was approved the previous year.

In the atmosphere of international peace and collaboration after the First World War, the IAU was founded in 1919, with the purpose of recovering the time lost by the war and coordinating the efforts of professional astronomers from all over the world. Its activity was divided into different committees, each one dedicated to a different field of astronomy, and among them was the lunar committee. Naturally, due to her encyclo-

---

[3] In particular, she studied the mysterious binary system of beta Lyrae, a puzzle that years ago, from the other side of the ocean, two other women, Antonia Maury and Henrietta Leavitt, had also tried to solve. It consists of two massive stars in close orbits, not only eclipsing each other, but transferring mass from one to the other.
[4] At that time, he was the secretary of the society. He had previously been its president, in the 1902–1904 period.

pedic work, Blagg was elected member of the newly founded committee, one of four women who were part of the initial cohort of the IAU.

But Mary Adela had a reserved nature, attending few meetings of the IAU and working mainly from home and by correspondence. She came to only two general meetings of the IAU, the first one held in 1925 in Cambridge, United Kingdom, when she was sixty-seven. She is likely one of the several women in that year's group photo of attendees (Figure 70), although there are quite a few women who could be in their sixties in the picture, and their names have not been recorded[5].

Despite her age, her subsequent work as a member of the IAU continued actively. At the next meeting of the IAU, which she also attended, in Leiden, Holland, in 1928, she presented a list of ambiguous lunar features not included in the previous catalog. The lunar commission decided that she and Karl Müller (1866–1942), an official of the Czech Government of Viennese origin and amateur astronomer without training, like Blagg (but, like her, patient and without fear of laborious work), would examine

Figure 70 IAU 1925 group photo, Cambridge, United Kingdom.

[5] Stearn & Sons (1925). *International Astronomical Union Cambridge 14th–22nd July 1925* [digital image]. http://www.dspace.cam.ac.uk/handle/1810/238507

lunar photographs to exclude doubtful objects from catalogs and pre-pare a definitive list.

The job took several years, during which the two were continuously in contact while working. Blagg got the best photographs from the obser-vatories in Paris, Mount Wilson, and elsewhere. Their final catalog, a work in two volumes called *Named Lunar Formations*, which listed some 6000 lunar features, with their coordinates and sizes, was the remarkable work of two amateurs who had not received any salary for it. It was presented at the IAU meeting of 1932 in Cambridge, Massachusetts. The lunar commission of the IAU also decided that the authors, in compen-sation, should have a lunar crater named after them. Both Blagg and Müller refused, but the commission overruled their demurral, and their names were added to the list published in 1935. This list of names thereby became internationally official, since the IAU is the body that has been delegated the right to name the features on the surface of the Moon and other bodies of the solar system[6].

Although Mary Adela arrived late to astronomy, her scientific career was long, she worked until the end of her life (when *Named Lunar Formations* was published, she was seventy-seven years young), and left an indelible mark on the history of science. The last years of her life she had heart problems, perhaps due to stress related to the Second World War, and she died in 1944. She did not witness the beginning of the space age, and how it was going to affect her subject with the revelation of the hidden face of the Moon: a whole new world to name.

In the same way that Mary Adela Blagg kept a low profile, the Blagg crater is the smallest of our women's collection, very round, almost without erosion (which indicates that it is relatively recent), and barely 5 km in diameter. It is located almost right in the center of the visible face of the Moon, very close to Bruce (and visible on the map at the end of the chapter on Bruce).

---

[6] If today you visit the website of the lunar nomenclature commission of the IAU (http://planetarynames.wr.usgs.gov/Page/MOON/target), you can access the list of names of all the craters and the date on which each name was accepted by the IAU. The oldest date from 1935.

17

Mary Proctor (1862–1957)

The sight is a revelation no words can adequately describe. One is speechless while regarding the wonders of this superb planet, encircled by the golden-hued rings shaded with a mere suggestion of purple tint. The faint outline of the planet was visible through the transparent crape ring, and the shadow of the planet could be seen on the ring itself.

PROCTOR about Saturn,
The New York Times
(August 27, 1911)

Mary Proctor was born in 1862 in Dublin, Ireland, the daughter of Richard Proctor (1837–1888) and his wife Mary. Richard was one of the best-known astronomers of the late nine-teenth century and an author of books and articles about astronomy for the public, recognized for this work by being elected to the Royal Astronomical Society in 1866. His daughter followed in his footsteps.

Figure 71 Mary Proctor. Digital Penn Library.

As a young girl, Mary helped her father with his manuscripts, taking great pride in the care of his library, and reading his books even before she could understand them. She graduated from *The College of Preceptors* (today *The College of Teachers*) in London. The family emigrated to the United States in 1882. She attended a course on astronomy at Columbia University in 1897.

Her father began the publication of a scientific journal entitled *Knowledge*, and Mary played an active role in its production. She wrote a series of articles on comparative mythology, and in 1895 published her first popular book on astronomy: *Stories of Starland*, intended for a young audience. Her clear writing earned her the respect of professional astronomers, and in due course she took a job as astronomy editor for *The*

The Women of the Moon. Daniel R. Altschuler Stern and Fernando J. Ballesteros Roselló.
© Daniel R. Altschuler Stern and Fernando J. Ballesteros Roselló 2019. Published in 2019
by Oxford University Press. DOI: 10.1093/oso/9780198844419.001.0001

Figure 72 Advertising brochure for the Stereopticon.

New York Times and began offering public lectures in New York, where she lived. She became a very successful lecturer (delivering over a thousand by 1910) and took singing lessons to improve her voice and delivery. In her lectures, she used the new "stereopticon," which made it possible to project images on a screen.

In 1913, she went on a lecture tour in New Zealand to promote the construction of a solar telescope. The *Dominion* of Wellington reported[1]:

> Miss Proctor's lecture needs little description, if indeed it were capable of such. It was similar to others she has given in the inimitable charm and freshness with which it was delivered, for Miss Proctor can make astronomy, the oldest of all sciences, as interesting as some new and very wonderful fairy tale. Last night, the mere listener without any special

---

[1] https://paperspast.natlib.govt.nz/newspapers/DOM19130502.2.70

knowledge of his own could not but be absorbed by Miss Proctor's references of the immensity of space and to the sketchy story she told of the growth of our planet – Earth. She spoke of the long almost infinite time when the Earth was in such a state that no life was possible upon it, and of the time of death which should eventually come when again life could not exist upon it; she spoke of the time when the world was formless, unshapen, and lifeless, of the time when wastes of water were its only surface, of the time when the oceans had subsided and the land was covered with tropical vegetation; of the ice age when the polar caps slid down and enveloped the earth in ice, of the age following, the springtime of the earth, the age of genesis, and of the time when there would be no water and no life. This was only one short passage of Miss Proctor's lecture, and all the rest of it was equally absorbing, and both educative and entertaining. [The admiration, perhaps even envy, of the authors is clear.]

She was elected in 1898 to the American Association for the Advancement of Science and in 1916 to the Royal Astronomical Society. Between 1895 and 1940 she authored 16 books on astronomical topics.

There is no better way to get acquainted with Mary Proctor than by reading some of her prose. On the occasion of the passage of Halley's Comet, in May 1910, she wrote an article in *The New York Times*. Spectroscopic studies of the comet had been made for the first time, indicating that it contained cyanogen $(CN)_2$, a toxic gas. The eclectic French astronomer Camille Flammarion (1842–1925) made inflammatory remarks to the effect that when the Earth passed through the comet's tail, (as it would do on May 18), "the gas would impregnate [the Earth's] atmosphere and possibly snuff out all life on the planet". Some people rushed to purchase gas masks and "comet pills." Mary begins her article[2] "Fears of the comet are foolish and ungrounded," as follows:

A dismal report is circulating to the effect that Halley's Comet is about to cause the destruction of our planet, and as we draw nearer the fateful date of May 18, a grave feeling of apprehension is excited in the minds of those who are very naturally afraid of something they cannot understand. Here is a gigantic monster in the sky, with a head over two hundred thousand miles in width, (according to measures recently made by Prof. Barnard of the Yerkes Observatory), and a train two million miles in length, rushing through space at the alarming rate of thousand miles a minute.

[2] http://query.nytimes.com/mem/archive-free/pdf?res=FA0D17F73E5D11738DDDA1 0894DD405B808DF1D3

On May 18 the earth will be plunged in this white-hot mass of glowing gas, and, according to the report of the ignorant and superstitious, the world will be set on fire.

These sensation makers further say that the oceans on the side facing the comet will be boiled by the intense heat, and the land scorched and blistered as the dread wanderer passes by on its baneful way.

After some examples of previous appearances of the comet, variously interpreted as a bad omen or a harbinger of good fortune, and commenting on the uniformly harmless passages of the past, she continues:

It would be well if our own times were free from these idle fears concerning cometary influence, for it would prove that men were unaffected by the debasing effects of ignorance and superstition. Why should bodies travelling uniformly in definite paths under the influence of the law of gravity be regarded as special messengers warning men either of good or evil approaching them? Good and evil prevail in the world, comet or no comet, but the broad shoulders, or rather head, of Halley's comet must bear the blame of every disaster likely to occur on or before the fateful May 18.

A man who could believe that Halley's comet, whose return was predicted within four weeks in 1759, and within three days, (so greatly has our knowledge of the planets and their disturbing influence on the comets increased,) in 1910, was a messenger specially sent from heaven on these occasions, or on the other occasions during which it has visited our neighborhood during the past 2000 years, would believe anything. In such a case reasoning is almost hopeless. Astronomers are being suspected as conspiring together to keep the uninitiated in ignorance the true fate awaiting our planet.

Mary ends her article with:

The poisonous cyanogen gas, which has been detected in the composition of the train, should in no way cause unnecessary alarm. Though the size of the comet is enormous, the particles of which it is made are excessively minute. As a result, according to Prof. Mitchell of Columbia University, the number of particles per cubic mile in the comet's tail is vanishingly small. Hence, though there may be some cyanogen gas in the tail, it is there in such small quantities that could we have a cubic mile of the tail concentrated into a glass beaker in the laboratory it would probably take the greatest refinement of chemical research to detect the cyanogen. In addition, the earth is covered with a shell of atmosphere thousands of

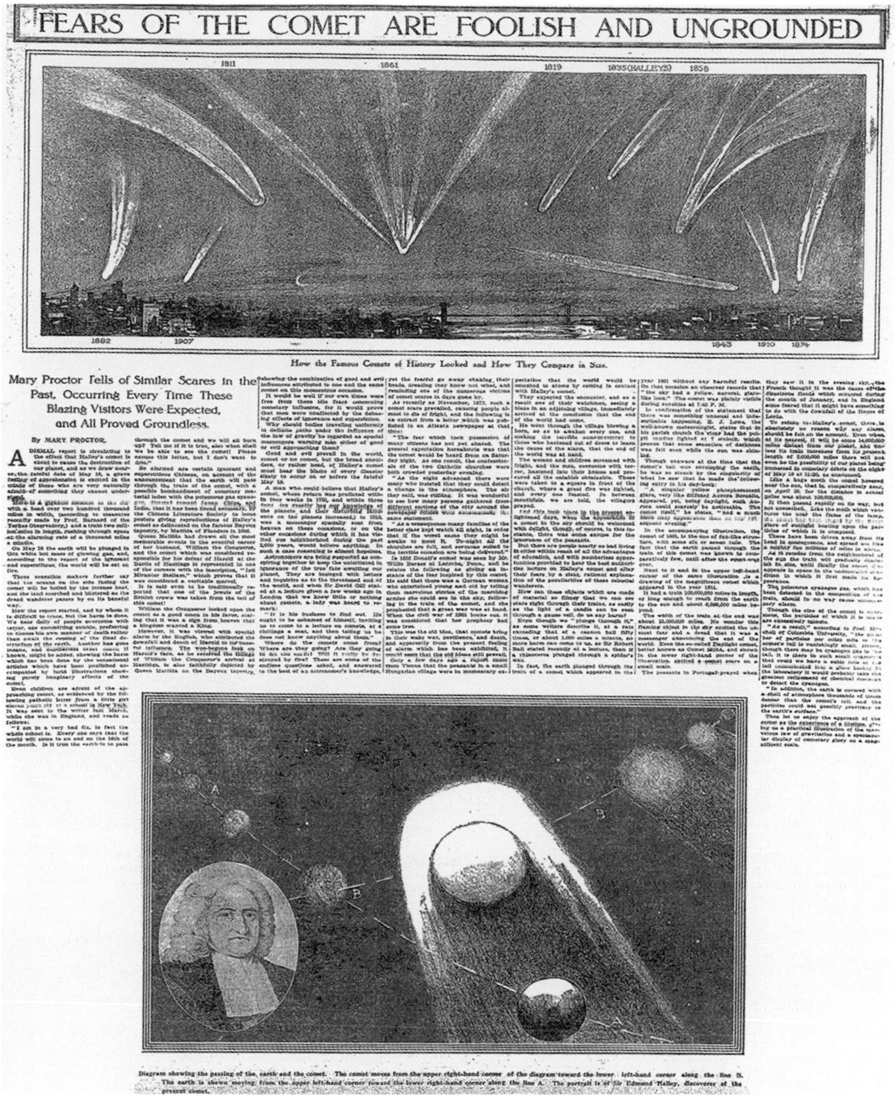

FEARS OF THE COMET ARE FOOLISH AND UNGROUNDED

How the Famous Comets of History Looked and How They Compare in Size.

Mary Proctor Tells of Similar Scares in the Past, Occurring Every Time These Blazing Visitors Were Expected, and All Proved Groundless.

By MARY PROCTOR.

Diagram showing the passing of the earth and the comet. The comet moves from the upper right-hand corner of the diagram toward the lower left-hand corner along the line B. The earth is shown moving from the upper left-hand corner toward the lower right-hand corner along the line A. The portrait is of Sir Edmund Halley, discoverer of the present comet.

Figure 73 The New York Times Sunday Magazine, May 8, 1910.

times denser than the comet's tail, and the particles could not possibly penetrate to the earth's surface.

Then let us enjoy the approach of the comet as the experience of a lifetime, giving us a practical illustration of the marvelous law of gravitation and spectacular display of cometary glory on a magnificent scale.

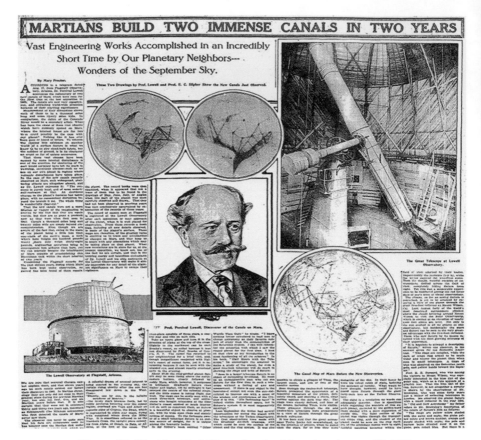

Figure 74 The New York Times Sunday Magazine, August 27, 1911.

Mary published this article more than a hundred years ago, but with minor changes, it could be published today, the next time a comet comes near Earth.

Another article—in part more credulous—was published by Mary on August 27, 1911. It carries the suggestive title: "Martians build two immense canals in two years." The idea of Martian canals had been set in motion after the Italian astronomer Giovanni Schiaparelli (1835–1910), observing Mars from the Milan Observatory in 1877 at opposition (when it is closest to Earth), reported linear features on its surface which he called *canali*. In Italian, *canali* means "channels", and can refer to a natural feature like a river channel, but something got lost in translation and the word became "canals" in English, implying an artificial origin.

Schiaparelli thought that these features were natural depressions in which water flowed to support Martian life.

Percival Lowell (1855–1916), a wealthy businessman, mathematician, and astronomer, speculated that the canals had been built by a Martian civilization to transport water from the poles to the temperate zones of Mars. In 1894, he built an observatory under the clear skies of Flagstaff, Arizona, to study Mars as an abode for life. Over many years, Lowell popularized the idea of Martians. The center of Mary's article is dominated by Lowell's face, surrounded by images of the observatory and drawings of Martian canals. The part of the article on Mars discusses changes observed in the canals as a result of Lowell's investigations (changes likely due to seasonal changes on Mars, atmospheric effects, and a bit of wishful thinking).

Proctor's article describes Lowell's findings, but dedicates its bulk to other astronomical topics, in particular to the planet Saturn. We suspect that she used the Martian theme just to draw attention, a clever strategy. She writes:

> A celestial drama of unusual interest is being enacted in the evening sky, the curtain rising shortly after the sun has disappeared low down in the West. Then − "Silently, one by one, in the infinite meadows of Heaven" − the lovely stars blossom forth, led by ruddy Arcturus hovering over the sunset region: overhead are Vega and Altair on opposite sides of Cygnus, the Swan, which is represented in olden star maps, flying with outspread wings down the Milky Way. It is sometimes referred to as the Northern Cross, the upright piece extending from Alpha, or Deneb, to Beta, or Albireo, at the foot of the cross. The cross-piece consists of three stars, a central star and one on each side.

> Take an opera glass and turn it in the direction of Alpha at the top of the cross. Slightly to the northwest can be seen Brook's comet, so called because it was just discovered by Prof. Brooks of Geneva, N.Y. The comet has returned to our neighborhood for a brief visit, and, though insignificant compared with Halley's Comet of recent fame, yet it is interesting because it can be seen with the unaided eye, and almost exactly overhead early in the evening.

> About 10:30 the ring-girdled planet Saturn may be seen rising above the northeastern horizon, closely followed by ruddy Mars, which however, it surpasses in brilliancy. [. . .]

> It has been said that the great astronomer Tycho Brahe felt overwhelmed, as in the presence of princes, when he gazed upon the stars, for to him they were monarchs of the sky. In their honor he wore his velvet robes of state, befitting the presence of royalty. What would he have

thought, however, could he have gazed upon Saturn through the great forty-inch lens at the Yerkes Observatory.

Wonders of the Sky is a book she wrote in 1931. Let's see what she tells us in Chapter II[3] about the moon, "a dead world":

The light of the moon is white, though at night when seen against the darkened sky it appears silvery in hue. Despite the seeming beauty of the moon, the Queen of Night has no light of her own. Her silver mantle is borrowed sunlight, for the moon is a dead world. There is not a sign of life, not a tree, a flower, or even a blade of grass to relieve the dreary monotony of boundless plains, mountain heights, or of yawning chasms. Not a sound could be heard on its surface; speech would be utterly useless, as there is no air to carry the sound of a voice. The silence of death must prevail. The absence of air must also produce startling effects in a lunar landscape, distant mountains appearing sharply outlined against a sky black as at night, their outlines only brought into sharp relief when illumined at sunrise. Equipped with a telescope, one can see these shadows slowly creeping across the plains, rugged precipices and deep sunken craters filled with shadow standing out in sharp relief against the regions gleaming in sunlight.

Could we transport ourselves to the moon, we should find no trace of water on its surface, nor the "seas" so fancifully named by Galileo, such as the Sea of Serenity, the Sea of Tranquility, the Bay of Rainbows, and others of like kind, names which are still retained on maps of the moon.

It is thought that this wonderful spirit died in London in 1957 at the age of ninety-six, but surprisingly and sadly, nothing is known of her final years.

Take your "opera glass" tonight and look for Proctor on the Moon. Her crater lies to the southeast of the nearby prominent crater Tycho. Proctor is an irregular, old crater probably caused by an oblique impact, 52 km in diameter, with walls that are 1300 meters high.

[3] Mary Proctor (1935). *Wonders of the sky*, p. 35. Frederick Warne and Co.

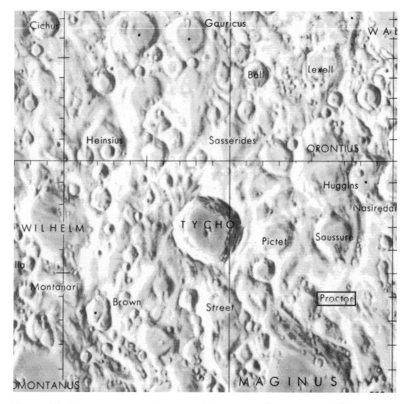

Figure 75 Location of crater Proctor. Courtesy of the Lunar and Planetary Institute, Houston, Texas.

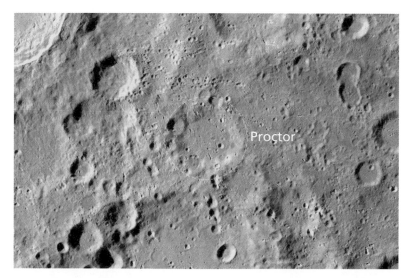

Figure 76 Lunar Reconnaissance Orbiter zoom on crater Proctor (image width is 150 miles).

18

Marie Skłodowska-Curie (1867–1934)

Figure 77 Marie Curie (photo taken 1907).

[F]rom this point of view, the atom of radium would be in a process of evolution, and we should be forced to abandon the theory of the invariability of atoms, which is at the foundation of modern chemistry[1].

It is hard to think that after so many centuries of development; the human race still does not know how to resolve difficulties in any way except by violence.[2]

MARIE CURIE

Of all the women of the Moon, Marie Curie is undoubtedly the best known, a legendary icon. Behind this image hides a heroic and tragic life, a life so full that this summary can only stimulate your curiosity to know more about this celebrated woman.

Marya Salomee Skłodowska (Manya to her friends and family) was born in Warsaw in 1867, the youngest of Vladislav and Bronislawa Skłodowska's five children. The Skłodowskas were both teachers, physics and mathematics for Vladislav, while Bronislawa, pianist and singer, was the director of a school for girls. These were very difficult times for the Poles: their country was divided between Russia and Prussia, both bent upon eradicating the Polish language and culture. These efforts hit the educational system particularly hard, as children were punished for speaking Polish, and in Warsaw, Polish teachers were replaced by Russians. As a consequence, Vladislav lost his job and the family was

[1] Radium and Radioactivity. By Mme. Sklodowska Curie, Discoverer of Radium from *Century Magazine* (January 1904), pp. 461–466. https://www.aip.org/history/exhibits/curie/article.htm

[2] Barbara Goldsmith (2004). *Obsessive genius: the inner world of Marie Curie*, p. 185. W.W. Norton.

The Women of the Moon. Daniel R. Altschuler Stern and Fernando J. Ballesteros Roselló.
© Daniel R. Altschuler Stern and Fernando J. Ballesteros Roselló 2019. Published in 2019 by Oxford University Press. DOI: 10.1093/oso/9780198844419.001.0001

driven into poverty. In the aftermath of the defeat of the Polish armed revolution of 1863, what resistance remained was intellectual, a network of educators who established a clandestine university whose aim was to rescue Polish culture and nationality. The stakes were high: Prussian Chancellor Otto von Bismarck (1815–1898) had declared[3]: "Hit the Poles until life fails, I have sympathy for their situation, but we cannot, if we want to survive, do anything but eliminate them, it is not the fault of the wolf that God created him as he is, and we nevertheless kill him when we can." (Charming, right?)

That was the environment in which Manya grew up. When she was eleven years old, both her older sister and her mother died, the former from typhoid fever and the latter from tuberculosis. She fell into a depression and declared that God did not exist. Her impoverished father did what he could, giving her what no one could take away: an education. She was a dedicated student, and at fifteen she graduated, best in her class. The university in Warsaw did not accept women and so, between 1885 and 1991, she worked to support her family as a governess in the home of the wealthy Zorawski family. While in their household, she had a romantic relation with the eldest son of the family, Kazimierz, who studied at the university, but his family was opposed and the relationship did not prosper. In 1887, she wrote to a cousin: "If men do not want to marry young poor women, let them go to the devil! Nobody is asking them for anything. But why do they offend by altering the peace of an innocent child?"

She also taught rural children to read and write Polish, a forbidden and risky activity. She had agreed with her older sister, Bronislawa Skłodowska (1865–1939), that with her savings she would help Bronya go to Paris— where women were accepted at the university—to study medicine, and later Bronya would help her in turn come to Paris, to study science.

But Manya was, for the moment, in Warsaw, and she fell into a severe depression (something that would happen many times during her hectic life). Despite her state of mind, she seized the scholarly opportunity that was available, and studied physics and chemistry informally, thanks to her cousin, Józef Boguski (1853–1933), who had been an assistant to the Russian chemist Dmitri Mendeleev (1834–1907) in St. Petersburg. Boguski directed a laboratory at the museum of industry and agriculture, enabling him to offer secret educational

---

[3] From a letter to his sister Malwine.

activities at night and on weekends, and allowing Manya to gain laboratory experience. He followed the Russian nihilists, and much later Marie would recall[4]: "You cannot hope to build a better world without improving the individual. To that end each of us must work on their own improvement, and at the same time share the general responsibility for all humanity, our particular obligation is to help those for whom we think we can be more useful," a motto that still rings true in the present.

In November 1891, the year of Sofia Kovalévskaya's death, Manya, twenty-four years old, took her few belongings and climbed on the train with the cheapest ticket for Paris to pursue her dream. After four days of travel in fourth class—she had to bring her own chair, and let's not even talk about sleeping—she arrived in Paris. She lived for a time with her sister Bronya, who had married a Polish exile, but then moved to the Latin Quarter, in order to be closer to the University of Paris, where she enrolled as "Marie." In that small room on the sixth floor, among prostitutes and frustrated artists, she endured the cold and ate poorly (so much so that her sister had to intervene and feed her on occasion), but she studied with fervor, and tried to catch up from the lost years of her education. Her arrival in Paris was something of a shock: she realized how much she lacked, including the fact that her French was not as good as she thought. Her tenacity and dedication paid off and she earned her Bachelor of Science degree in 1893, first in her class, and her Bachelor of Mathematics in 1894, second in her class. The question was what to do to continue her studies.

Around this time, she was invited to the quiet boardinghouse of a Polish friend, Józef Kowalski (1866–1927), Professor of Physics at the University of Freiburg, who was visiting Paris. There she met a young physicist who worked at the *École Supérieure de Physique et de Chimie Industrielles de la Ville de Paris* (ESPCI), who had been invited over by Kowalski to give Marie some advice. Pierre Curie was known for his discovery in 1880, with his brother Jacques, of the piezoelectric effect in crystals (the relationship between electricity and pressure), and he was an expert on magnetism. Marie writes in her autobiographical notes[5]:

[4] Eve Curie (1937). *Madame Curie*, p. 53. Da Capo Press.
[5] Marie Curie (1923, 2012). *Pierre Curie: with autobiographical notes by Marie Curie*, p. 34. Dover Publications.

As I entered the room, Pierre Curie was standing in the recess of a French window opening on a balcony. He seemed to me very young, though he was at that time thirty-five years old. I was struck by the open expression of his face and by the slight suggestion of detachment in his whole attitude. His speech, rather slow and deliberate, his simplicity, and his smile, at once grave and youthful, inspired confidence. We began a conversation which soon became friendly. It first concerned certain scientific matters about which I was very glad to be able to ask his opinion. Then we discussed certain social and humanitarian subjects which interested us both. There was, between his conceptions and mine, despite the difference between our native countries, a surprising kinship, no doubt attributable to a certain likeness in the moral atmosphere in which we were both raised by our families.

It was love at first sight—a bond both emotional and intellectual. Encouraged by Marie, Pierre completed his doctoral thesis for the Sorbonne—a study of the magnetic properties of materials as a function of temperature. In July 1895, Marie and Pierre married in a simple civil ceremony in the house of Pierre's parents, after which, aboard bicycles, they took their honeymoon.

These were crucial years in science. In 1895, Wilhelm Röntgen discovered the mysterious X-rays (the "X" itself representing a mystery—the unknown in an equation then as now), and in 1896, Henri Becquerel discovered another mysterious ray: the radioactivity (a term coined by the Curies) produced by uranium salts. X-rays were of great public and scientific interest, and soon were used to locate metallic objects in human bodies, something that would prove all too useful in the coming war. In contrast, Becquerel's rays did not produce such interest and he himself did not continue with these studies for long.

Marie and Pierre were very happy. In her biography[6], Eve Curie, her second daughter, born in December of 1904, tells us:

> During those happy days was formed one of the finest bonds that ever united man and a woman. Two hearts beat together, two bodies were united, and two minds of genius learned to think together. Marie could have married no other than this great physicist, than this wise and noble man. Pierre could have married no woman other that this fair, tender Polish girl, who could be childish or transcendent within the same few moments, for she was a friend and a wife, a lover and a scientist.

---

[6] Ève Curie (1937). *Madame Curie, a biography*, p. 141. Da Capo Press.

Figure 78 Portrait of Marie Curie and Pierre Curie about 1905.

Back in Paris, the Curies settled in a modest apartment, and when first daughter Irène was born (1897–1956), Marie faced the difficulties of life as a young scientist and mother—still a point where the science community loses too many talented women. In a silver lining to the death of Pierre's mother, his father, doctor Eugène Curie, moved to live with them, to help with the budget and help taking care of his grand-daughter. At the end of that year, Marie began her doctoral research. Looking for the newest and most interesting question, and with the advice of Pierre, she decided to study the strange radiation discovered by Becquerel. Using an electrometer designed by Pierre, very sensitive but difficult to operate, she began to measure the radioactivity produced by different minerals, working in a warehouse that had been assigned to them by the ESPCI as a laboratory. She soon discovered that thorium also emitted radiation. But more surprising was her finding that residues of pitchblende ore from which the uranium had been extracted produced more intense radiation than the uranium itself.

Pitchblende, also known as uraninite, contains uranium oxides, from which uranium is extracted. This metal, with atomic weight 89, had been discovered in 1789 by the German chemist Martin Heinrich Klaproth (1743–1817), who studied the composition of pitchblende. He called it "uranium" to honor the discovery of Uranus in 1781 by William Herschel, brother of our Caroline, as we have seen. After the process of extracting the uranium from the pitchblende (uranium was used as a pigment for glass and ceramics), the remaining waste material was discarded at the beginning of the twentieth century. But uraninite minerals—as Marie was discovering—also contain very small amounts of other highly radioactive elements that are the product of the radioactive decay of uranium, and which remain in extremely small concentrations in the residue. The Curies obtained from the Austrian government, owner of a uranium mine, a ton of waste and paid for its transport to France.

Separating that tiny fraction, studying its properties, and thus discovering two new elements, radium (Ra-88) and polonium (Po-84), was the great feat of the Curies, working for several years in a shed by the École de Physique. A ton of pitchblende contains less than one gram of radium, and over these years they processed several tons of that material—hard labor as much akin to working in a metal factory as a lab. They were joined by their friend André-Louis Debierne (1874–1949), who worked a day job as an assistant in a laboratory at the Sorbonne but after work came to the Curie's laboratory to help with the chemical separation. He collaborated with them the rest of his life, discovering the element actinium (Ac-89) in 1899.

In the journal of the French Academy of Sciences for July 1898, we read for the first time of polonium, in a communication from the Curies presented by Becquerel[7]: "We believe that the substance we have extracted from the pitchblende contains a metal not yet observed, related to bismuth from its analytical properties. If the existence of this new metal is confirmed, we propose to call it polonium, from the name of the country of origin of one of us." (Many other known elements—ruthenium, gallium, scandium and germanium—had also been named after the countries of their discoverers.) In December of 1898, the Curies published the discovery of radium. In that article we read[8]: "The various reasons that we have enumerated lead us to believe that the new

---

[7] Pierre Curie and Marie Curie (1898). Sur une substance nouvelle radioactive, contenue dans la pechblende. *Comptes rendus de l'Académie des Sciences*, Vol. 127, p. 175.

[8] Pierre Curie, Marie Curie, and G. Bémont (1898). Sur une nouvelle substance fortement radio-active, contenue dans la pechblende. *Comptes rendus de l'Académie des Sciences*, Vol. 127, p. 1215.

radioactive substance contains a new element to which we propose to give it the name of radium."

The radioactive products obtained by the Curies were handled without any precautions, and Marie's fingers were burned and her skin cracked. They breathed radon—a gas produced by radium—and ate in the laboratory; their notebooks remain dangerously radioactive to the present. Pierre walked around with a test tube of radium salts in his pocket, to show his friends the blue light emitted by the magical compound. Not knowing the nature of radiation and its biological hazards, they did not take special care when handling the radioactive materials. Soon Pierre noticed the burns left on his skin, so he stopped carrying the test tube; thus was born the idea of the biological effects of radiation, and the possible medical use of radioisotopes.

In June 1903, Marie defended her thesis before a jury presided over by Gabriel Lippmann (1845–1921; Nobel Prize in Physics in 1908) and became the first woman to obtain a PhD in France. At the defense were her husband, Pierre, and her sister Bronya (who had come from Poland filled with pride); the physicists and friends Jean Perrin (1870–1942, Nobel Prize in Physics in 1926) and Paul Langevin (1872–1946); her father-in-law, Eugène Curie; and many other friends and colleagues who were familiar with Marie's discoveries. After the question session and its explanations, the president of the jury solemnly pronounced: "The University of Paris confers on you the title of doctor in physical sciences with the mention of tres honorable (very honorable). On behalf of the jury, Madame, I wish to express our congratulations." Manya had achieved her goal!

The most surprising thing is that this research was carried out in a "laboratory" that hardly deserved such a designation. The German chemist Wilhelm Ostwald (1853–1932, Nobel Prize in Chemistry in 1909) visited it and commented[9]: "It was a cross between a stable and a potato cellar, and, if I had not seen the work table with the chemical apparatus, I would have thought it a practical joke."

In 1903, the Royal Swedish Academy of Sciences awarded the Nobel Prize in Physics to Pierre Curie, Marie Curie, and Henri Becquerel "in recognition of the extraordinary services they have rendered by their joint researches on the radiation phenomena discovered by Professor Henri Becquerel". Initially the committee was going to honor only Pierre and Henri, but Magnus Gösta Mittag-Leffler, whom we already met in relation

[9] Sharon Bertsch McGrayne (2001). *Nobel Prize women in science: their lives, struggles, and momentous discoveries*, p. 23. Joseph Henry Press.

to Kovalévskaya and Leavitt, alerted Pierre Curie, who protested, and declared that he would not accept it if Marie was not included, and got the committee to include her. Thus, she became the first woman to win the Nobel Prize, and for thirty-two years, the only one (with a second prize in 1911), until her daughter Irène was awarded it in 1935.

After the Nobel Prize in 1903, the couple became celebrities, a situation that was not to their taste. But at the same time, this opened doors and finally, after several setbacks and misunderstandings, Pierre was offered a faculty position at La Sorbonne and a laboratory with three assistants; Marie was appointed as research director. He began to study the possible medical uses of radium. He had observed the death of laboratory animals exposed to radiation, but apparently, he did not realize that the deteriorating health and fatigue both the Curies themselves felt were due to the radiation they had been exposed to for years.

Extracted from pitchblende, radium was extremely difficult to obtain. In 1904, a gram of radium had a value of about $100,000. The medical applications of radium aroused the interest of industry, as metals companies vied to meet the now-exploding market demand for the material. Pierre and Marie, after considering the requests that they were receiving to publish the details of the process they had used to extract radium from pitchblende, decided not to apply for a patent for the process, something that would have delivered them great wealth. They thought that their discoveries belonged to all humanity—how nice it would be if more researchers felt this way!

On Thursday, April 19, 1906, Pierre had lunch with some professors of science from the faculty. At two thirty in the afternoon he said goodbye to his colleagues, and with his umbrella open, since it was raining, he headed for the Seine, to the office of the publishers of *Comptes Rendues* in Gauthier-Villars, to correct some galleys. He found that Gauthier-Villars had closed because of a strike, and returned through the busy streets of Paris. He limped, with his leg bones damaged by exposure to radiation. When crossing the Rue Dauphine he did not notice a heavy carriage carrying military uniforms and fell between the legs of the horses; alas, the big rear wheel of the carriage crushed his head. Nothing could be done by those who tried to assist him at a nearby police station. Pierre was dead.

Marie lost her best friend, her scientific colleague, and her beloved husband. In her words[10]: "so perished the hope founded on the wonderful

---

[10] Marie Curie (1923). *Pierre Curie*, p. 66. Dover Publishing.

being who thus ceased to be" With his death, her full life became a sleepwalking life. Her daughter Ève writes: "From the moment when those three words, 'Pierre is dead,' reached her consciousness on that day in April, Madame Curie became not only a widow, but at the same time a pitiful and incurably lonely woman." She wrote: "I endure life, but . . . never again will I be able to enjoy it . . . I will never be able to laugh genuinely until the end of my days." And so it was. Pierre was buried in the family tomb in Sceaux. You can sense Marie's sorrow in the image at the start of this chapter.

The University of Paris offered Pierre's position to Marie, and she thereby became the first woman in the 650-history of the Sorbonne to

Figure 79 Marie with Eve (left) and Irène in 1908.

occupy a faculty position, and the second European woman to obtain a chair—seventeen years after the first one: Kovalévskaya. When she entered the room, on November 5, 1906, to offer her first lecture, she found herself facing an audience of hundreds of people—journalists, elegantly dressed women, students, and onlookers, who applauded for five minutes before she could begin.

In her diary she wrote:

> Yesterday I gave the first class to replace my Pierre. What a pity and what despair! You would have been happy to see me as professor at the Sorbonne, and I myself would have done so for you. But to do it in your place, O Pierre, could one dream something more cruel, and how I suffered, and how discouraged I feel. I feel that any faculty of living is dead in me, and I have only the duty to raise my children and also the will to continue the accepted task. Perhaps also the desire to prove to the world and especially to myself that the one you loved so much really had some value.

An addition to teaching, Marie had research commitments. The complicated nature of the process to obtain polonium and radium made replication by others difficult and there remained doubters as to whether or not they were in fact new chemical elements. The 1903 Nobel Prize, recall, was awarded to the Curies not for discovering new elements but "in recognition of the extraordinary services they have rendered by their joint researches on the radiation phenomena." One who doubted was the eminent William Thomson, 1st Baron Kelvin (1824–1907). Not knowing of the energy from radioactivity, he had calculated that the Earth could not be more than about ten million years old, a time too short for Darwin's biological evolution to operate.

He thought that the atom was indivisible and that, therefore, radioactivity had another explanation. His skepticism occasionally came across as obstinacy. Ernest Rutherford (1871–1937), one of the pioneers of radioactivity and Nobel Prize in Chemistry for 1908 for his research into the decay of atoms (his important research on the structure of the atom came later), commented in a letter to his wife: "Lord Kelvin has talked radium most of the day, and I admire his confidence in talking about a subject of which he has taken the trouble to learn so little. I showed him and the ladies some experiments this evening, and he was tremendously delighted and has gone to bed happy with a few small phosphorescent things I gave him."

Without Pierre, Marie spent four dogged years of hard work to show Kelvin—and the world—that radium was indeed a new element, and that radioactivity was a property of compound atoms. She succeeded in separating from tons of pitchblende a tiny amount of pure radium.

In 1907, she received a request from an Austrian woman to work as an assistant in her laboratory. The woman had experimented with radium salts, a gift from the Curies to the University of Vienna, in gratitude for the pitchblende obtained from Austria. The request was rejected and Lise Meitner (whom we will meet in the next chapter) ended up working in Berlin.

Personally, these were not happy years for Marie. In 1910, spewed by the tabloids dedicated to scandals between celebrities (and catering to the political right), the news exploded that Marie had a love affair with the physicist Paul Langevin, Pierre's student. The vicious attacks described her as a Polish foreigner, home wrecker, and, in those times of casual anti-Semitism, Jewish (she was not). The supposed affair became the most sensational event since the theft of the Mona Lisa. The university considered removing her from her position and the matter nearly reached the buffoonery of a duel between a journalist and Langevin that, fortunately, never took place. No one thought to criticize Langevin for having a lover. As if this weren't enough, at the beginning of 1911 the Académie des Sciences Française considered and rejected the candidacy of Marie (only recently, in 1979, was the first woman elected: Yvonne Choquet-Bruhat).

Years later, Langevin had a daughter with one of his ex-students and asked Marie to employ her in her laboratory. She agreed. The circle was closed when the granddaughter of Marie, daughter of Irène, Hélène Joliot (1927–) and Langevin's grandson, Michel Langevin (1926–1985), got married. Continuing the family business, as it were, both worked in the field of nuclear physics.

In the midst of all this mess, Marie learned in 1911 while attending the first Solvay conference in Belgium, that she was the sole winner of a second Nobel Prize, this one in Chemistry "in recognition of her services to the advancement of chemistry by the discovery of the elements radium and polonium, by the isolation of radium and the study of the nature and compounds of this remarkable element."

Although weak and ill, Marie went to Stockholm to receive the award, accompanied by her sister Bronya and her daughter Irène, who observed the solemn ceremony. Twenty-four years later (a year after Marie's death), Irène and her husband would be the laureates. In Marie's

Figure 80 One among many. Solvay congress 1911. Marie surrounded by men in black. Standing from left to right: O. Goldschmidt, M. Planck, H. Rubens, A. Sommerfeld, T. Lindemann, M. de Broglie, M. Knudsen, F. Hasenöhrl, H. Hostelet, T. Herzen, J. Jeans, E. Rutherford, H. Kamerlingh Onnes, A. Einstein, P. Langevin. Sitting from left to right: W. Nernst, M. Brillouin, E. Solvay, H. Lorentz, E. Warburg, J. Perrin, W. Wien, M. Curie, H. Poincaré. Courtesy of the International Solvay Institutes, Brussels.

acceptance speech she did not forget her beloved Pierre, saying: "Before broaching the subject of this lecture, I should like to recall that the discoveries of radium and of polonium were made by Pierre Curie in collaboration with me. We are also indebted to Pierre Curie for basic research in the field of radioactivity, which has been carried out either alone, or in collaboration with his pupils."

Soon radium, with its mysterious energy and its ghostly radiance, acquired amazing powers in the public imagination. Novel and fictitious uses were invented. Glowing paints prepared with radium salts and a radioluminescent compound were applied to all types of instrument dials, including novel clocks that could be read in the dark. The cosmetics industry invented all kinds of creams and elixirs that supposedly contained radium and even more supposedly produced miracles of beauty and health. In Germany, in the 1930s, *Radium Schokolade* could be bought, with rejuvenating powers. The ad said: "The chocolate to eat and drink is prepared with delicious and ripe cocoa seeds. The radium is added in such a way that it does not affect the taste."

Many women were employed by the watch industry to apply luminous paint with a paintbrush to the numbers on the dials. To keep the tip of the brush sharp they placed it in their mouth, and many stained their teeth or lips with paint. In a cruel twist, radium has chemical properties similar to calcium, so if it is ingested it is incorporated into the bones, where the radioactivity does its damage. The case of the *radium girls* is well known. They developed anemia, osteonecrosis of the jaw, and cancer by exposure to radium. Some women had ingested so much radium that, even today, radioactivity is detected in their graves. Survivors sued the United States Radium Corporation for damages, and after many years of litigation, they managed to settle out of court. It is a key case in the early advent of occupational safety laws. In 1928, Dr. Sabin Arnold von Sochocky, inventor of the paint and co-founder of the company, died of aplastic anemia, victim of his own invention. Since the half-life of radium is 1600 years, the dials of these old instruments are still radioactive and potentially dangerous, although they are no longer luminous (due to degradation of the radioluminescent compound).

By mid 1914, a Curie dream was fulfilled. After several years of collaboration between the university and the Pasteur Institute, the construction of the *Institut du Radium* (located on Rue Pierre et Marie Curie) was completed. Today it is one of the most advanced research centers in biophysics, cell biology, and oncology. The radioactivity laboratory would be led by Marie, and an adjacent laboratory to study the therapeutic applications by the eminent physician Claudius Regaud (1870–1940).

The First World War soon drew the world's attention from basic science. Marie, convinced that X-rays could save many lives by facilitating the location of metal fragments in the bodies of wounded soldiers, organized a mobile X-ray service. She spent two months seeking out equipment in laboratories, garnering support from people with resources and cars, and wrangling with the military authorities to allow her to move to the front lines. Marie, almost fifty years old, found herself at the wheel of a Renault truck, equipped with an electricity generator and an X-ray machine. Under the skeptical gaze of the military, she began to travel to the war front to treat the wounded, and in November she allowed young Irène (by then sixteen) to join her efforts. Over the course of the war years, she trained dozens of nurses and technicians. At war's end, four years later (a butchery that cost the lives of some fifteen million soldiers and civilians, not counting the enormous number of wounded), she had installed some two hundred radiography stations, and over one million soldiers had been examined. Marie and Irène were exposed to high doses of radiation, with deleterious effects on their health. The French government never recognized what they had done.

After the war, Marie dedicated herself to organizing her institute, but devastated France did not have the resources or the will to do much.

In 1921, accompanied by her two daughters, she arrived in New York on her first visit to the United States. She was welcomed as an international celebrity, including a front page in *The New York Times* that proclaimed: "Madame Curie plans to cure all cancers." She had to bandage her hand, hurt by the many handshakes she received. She was entertained by President Warren Harding (1865–1923), who gave her a gift of great value: a gram of radium, purchased with the proceeds of a fund drive organized by Marie Mattingly Meloney (1871–1943)[11].

The inspiration for all this was the promise to treat and cure cancer. Harding held forth at the ceremony:

> We greet you as foremost among scientists in the age of science, as leader among women in the generation which sees woman come tardily into her own [. . .] In testimony of the affection of the American people, of their confidence in your scientific work: and of their earnest wish that your genius and energy may receive all encouragement to carry forward your efforts for the advance of science and conquest of disease: I have been commissioned to present to you this little vial of radium. To you we owe our knowledge and possession of it: and so, to you we give it, confident that in your possession it will be the means further to unveil the fascinating secrets of nature, to widen the field of useful knowledge, to alleviate suffering among the children of man. It betokens the affection of one great people to another.[12] [Of course, the real vial was already deposited elsewhere.]

Marie replied: "I accept this rare gift Mr. President, with the hope that I may make it serve mankind. I thank your countrywomen in the name of

[11] "Missy" Meloney was a friend and admirer of Marie and editor of *The Delineator*, a women's magazine.
[12] *The New York Times*, 1921, May 21.

France. I thank them in the name of humanity which we all wish so much to make happier I love you all: my American friends, very much"[13].

She devoted her time to raising funds and converting the institute into one of the most important in the area of nuclear physics. There, in 1934, her daughter Irène, together with her husband, Jean Frédéric Joliot, achieved the transmutation of one element into another (the dream of the alchemists), and in 1939, Marguerite Perey (1909–1975), a disciple of Marie, discovered francium (neither of these women was honored with a lunar crater). During the postwar period, she received numerous honors, invitations to inaugurate "Curie institutes" in several countries, and participate in scientific congresses. She visited the United States for a second time in 1929, weak, with vision problems and anemia (as a consequence of irradiation), just before the fateful black Tuesday that marked the Great Depression. President Herbert Hoover (1874–1964) received her in the building of the National Academy of Sciences and gave her a check for $30,000 to buy a gram of radium for a new institute that Marie, along with her sister Bronya, was promoting in Warsaw. It was inaugurated in 1932, with Bronislawa as the first director.

In 1933, she participated in a meeting in Madrid outside of her scientific field. The Committee of Letters and Art of the League of Nations was convened to discuss the "future of culture," an event at which some participants, including Paul Valéry, Gregorio Marañón, Salvador de Madariaga, and Miguel de Unamuno, blamed science for the "crisis of culture." In Marie's presentation, she explained:

> I am among those who think that science has great beauty. A scientist in his laboratory is not a mere technician: he is also a child confronting natural phenomena that impress him as though they were fairy tales. We should not allow it to be believed that all scientific progress can be reduced to mechanisms, machines, gearings, even though such machinery also has its beauty.
>
> Neither do I believe that the spirit of adventure runs any risk of disappearing in our world. If I see anything vital around me, it is precisely that spirit of adventure, which seems indestructible and is akin to curiosity.

Weakened by years of exposure to radiation, Manya died at the age of sixty-seven of leukemia (a fate that would befall her daughter Irène as well) on July 4, 1934 and was buried, in a simple ceremony attended by family and friends in a tomb next to Pierre's in Sceaux. Her brother,

---

[13] *Science*, 1921, p. 497.

Józef Skłodowski (1863–1937) and her sister Bronya, who did not arrive in time to see her still alive, threw handfuls of earth on her coffin that they had brought from Poland. Her faithful friend Debierne went on to direct the institute.

In a ceremony in her memory held in New York in 1935, Einstein spoke with feeling of Marie, her humanity as much as her scientific work[14]:

> At a time when a towering personality like Mme. Curie has come to the end of her life, let us not merely rest content with recalling what she has given to mankind in the fruits of her work. It is the moral qualities of its leading personalities that are perhaps of even greater significance for a generation and for the course of history than purely intellectual accomplishments. Even these latter are, to a far greater degree than is commonly credited, dependent on the stature of character. It was my good fortune to be linked with Mme. Curie through twenty years of sublime and unclouded friendship. I came to admire her human grandeur to an ever-growing degree. Her strength, her purity of will, her austerity toward herself, her objectivity, her incorruptible judgment—all these were of a kind seldom found joined in a single individual. She felt herself at every moment to be a servant of society, and her profound modesty never left any room for complacency. She was oppressed by an abiding sense for the asperities and inequities of society. This is what gave her that severe outward aspect, so easily misinterpreted by those who were not close to her—a curious severity unrelieved by any artistic strain. Once she had recognized a certain way as the right one, she pursued it without compromise and with extreme tenacity.

Sixty years later, her remains were moved, along with those of Pierre, to the Panthéon in Paris to rest together with 76 men. In François Mitterrand's speech as President of the French Republic at the solemn occasion, on April 20, 1995, before a large audience which included the President of Poland, Lech Wałęsa, and addressing especially the grandchildren and great-grandchildren of Marie, he stressed that she had been the first French woman to be a doctor of science, a professor at the Sorbonne, and also to receive a Nobel Prize, and that she was also the first woman to rest in the famous Pantheon on her own merits (she was the only one, until in 2014 Germaine Tillion and Genevieve de Gaulle-Anthonioz, a niece of former President Gen. Charles de Gaulle, both

---

[14] Statement for the Curie Memorial Celebration, Roerich Museum, New York, November 23, 1935. Published in *Out of My Later Years*, New York: Philosophical Library, 1950.

Nazi resistors, were interred there. In 2017, their remains were joined by a fourth woman, Simone Veil).

He also said: "But there is another symbol that this night draws to the attention of the Nation before which I have the honor to speak: that of the exemplary struggle of a woman who has decided to impose her abilities on a society that reserves to men the intellectual functions and responsibilities."

Ironically, behind him, one could read the pantheon's inscription: "*Aux grands hommes . . .*"

Asteroid 7000 Curie, discovered on November 6, 1939, bears her name. On the Moon, crater Sklodowska is next to Curie, on the far side. It is the largest one for our women, with a diameter of 127 km.

Figure 81 Photo by one of the authors.

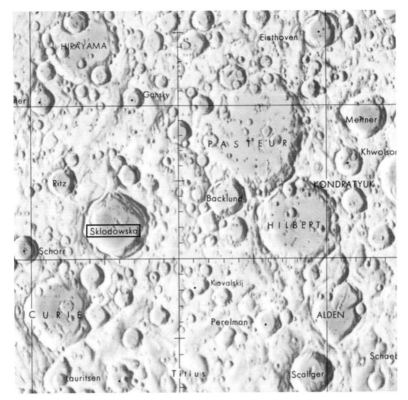

Figure 82 Location of crater Sklodowska. Courtesy of the Lunar and Planetary Institute, Houston, Texas.

Figure 83 Lunar Reconnaissance Orbiter zoom on crater Sklodowska (image width is 150 miles).

19

Lise Meitner (1878–1968)

Figure 84 Lise Meitner (photo taken 1906).

I must stress that I myself have not in any way worked on the smashing of the atom with the idea of producing death-dealing weapons. You must not blame us scientists for the use to which war technicians have put our discoveries.

> Response to an invitation (1943) to work with OTTO ROBERT FRISCH and other British scientists at LOS ALAMOS, with the Manhattan project

Lise Meitner[1] was born in Vienna in 1878, the third of eight children of Hedwig and Philipp Meitner, a Jewish, middle class, progressive, and one might say enlightened family. Vienna, on the banks of the Danube, was a dynamic city, full of immigrants of different ethnicities—a city whose dynamism had a hard edge that bordered on chaos, with social problems of all kinds. From this breeding ground would emerge Sigmund Freud (1889–1945), Ludwig Boltzmann (1844–1906), Erwin Rudolf Josef Alexander Schrödinger (1887–1961), Theodor Herzl (1860–1904), and (unfortunately) Adolf Hitler (1889–1945) among others.

After completing the curriculum at a women's high school with the reduced education that the institution considered sufficient for a woman, her options at fourteen were few. She opted for a private school to specialize as a French teacher. In a 1951 letter, Meitner would write[2]:

[1] Ruth Lewin Sime (1996). *Lise Meitner: a life in physics*. University of California Press.
[2] Ruth Lewin Sime (1996). *Lise Meitner: a life in physics*, p. 7. University of California Press.

The Women of the Moon. Daniel R. Altschuler Stern and Fernando J. Ballesteros Roselló.
© Daniel R. Altschuler Stern and Fernando J. Ballesteros Roselló 2019. Published in 2019 by Oxford University Press. DOI: 10.1093/oso/9780198844419.001.0001

"Thinking about the time of my youth . . . one realizes with some amazement how many problems were presented in the life of a young woman, problems that now seem unimaginable. Among the most difficult was the possibility of a normal intellectual training."

In 1899, the University of Vienna began admitting women, and Lise, at the age of twenty-three, after preparing for two years (to cover the eight that were sorely lacking), managed to pass the entrance exam (to give a sense of the difficulty of the "matura", only four of fourteen takers passed). She entered the University of Vienna in 1901. She attended the physics courses of one of the most eminent physicists and respected teachers of the time: Ludwig Boltzmann (1844–1906). Boltzmann is recognized as a pioneer in atomic theory (not all physicists at the time accepted the reality of atoms), and in particular for his statistical explanation of the second law of thermodynamics (Boltzmann has a crater on the Moon 76 km in diameter). Lise earned her doctorate in 1906, at twenty-eight.

She spent a year in Vienna collaborating with Stefan Meyer (1872–1949) to understand the characteristics of a new phenomenon: radioactivity; they were starting to characterize the emissions of the new radioactive elements (those radium salts, a gift from the Curies), which, at this early stage of understanding their nature, were known as alpha radiation (we now know that they are helium nuclei), beta radiation (electrons), and gamma radiation (high-energy photons). Meitner published her results on the dispersion of alpha particles in the prestigious journal *Physikalische Zeitschrift*.

Her interest in radioactivity led her to apply to work at Curie's new laboratory in Paris, under the recent Nobel Prize winner. She received a refusal that would eventually make her the most important scientific rival of Marie and Pierre's daughter, Irène Curie.

In 1907, with the approval of her parents, she moved to Berlin, at that time the Mecca of physics; she thought that she would remain there for a year to improve her knowledge, without imagining that she would stay for thirty years, before being forced into exile. In Berlin she met the chemist Otto Hahn (1879–1968), the same age as her, with whom she established a long collaboration and friendship. They formed a good team; he was a careful and methodical chemist, she, a brilliant physicist suggesting new experiments. Together with Hahn she worked on experiments to elucidate the nature of the new radioactive elements, at a time when the structure of the atom was not yet understood, nor even the fact that the chemical elements were not distinguished by their mass, but by their electrical nuclear charge (the number of protons), the discovery which led to the idea of isotopes of an element. Meitner and Hahn began producing interesting results and published numerous articles in the *Physikalische Zeitschrift*.

Gender discrimination appeared on many sides. Women were not allowed by Prussian law at the chemistry institute, and while its director, Emil Fischer (1852–1919, Nobel Prize in Chemistry, 1902), kindly allowed her to work in a laboratory in the basement, she was forbidden to go to the other laboratories. A year later the laws were changed and women were legally admitted to universities.

Such affronts could be personal as well as institutional: in 1908, Ernest Rutherford visited Berlin with his wife after receiving the Nobel Prize in Chemistry. While Hahn and Rutherford met to discuss science, Lise had to accompany Mrs. Rutherford for Christmas shopping. We can imagine how she must have felt about this.

The editor of the important *Brockhaus* encyclopedia, after reading some magazine articles written by "L. Meitner" about radioactivity, asked the magazine's director for the address of *Herr* Meitner to request a contribution on radioactivity. When they clarified that it was *Fräulein* Meitner, he replied that "it would not occur to me to publish an article written by a woman." In 1922, already famous, she gave a talk on "the meaning of radioactivity for cosmic processes," and a journalist in his review of the event wrote that it was about "the meaning of radioactivity for cosmetic processes."

Despite these circumstances—including the fact that without a formal position she was still forced to live frugally on what her father sent her—discrimination could not dim her brilliance. She formed deep friendships with eminent physicists, in particular James Franck (1882–1964, Nobel Prize in Physics in 1925 with Hertz) and Max Planck (1858–1947, Nobel Prize in Physics, 1908). Albert Einstein (1879–1955) was also a friend who referred to her as "our Marie Curie."

The *Kaiser-Wilhelm-Institut für Chemie* (KWI) was inaugurated in Dahlem, Berlin in 1912. Otto Hahn was offered a position as a member of the institute and professor, with an annual salary of 5000 marks. Meitner, meanwhile, was accepted as a visitor without pay. Things improved somewhat when that same year Planck named her his assistant, the first woman so appointed, gaining her some official recognition and her first proper research salary. Then Fisher gave her a position equal to Hahn's, but with less salary. Two years later, at the beginning of the war in 1914, she got a raise to 3000 marks.

The war brought about a radical change in the KWI; some, like Hahn, Franck, Gustav Ludwig Hertz (1887–1975), and Hans Geiger (1882–1945), marched into military service expecting a short and victorious war; and others, like Fritz Haber (Nobel Prize in Chemistry, 1918), dedicated their efforts to finding alternative materials for the arms industry—affected by blockades and embargoes—and the development of chemical weapons.

In July 1915, after learning that Marie Curie and her daughter Irène worked with radiological equipment in hospitals behind the French front, Lise decided to join the Austrian army as a nurse and X-ray technician, and went to work at the Russian front. After horrible experiences of war in several theaters, tired and disillusioned, she returned to the KWI at the end of 1916.

In 1917, Fisher named her director of the Physics section of the KWI, with an increase in salary. While Hahn was at the war front, Lise continued with a series of difficult experiments that finally culminated with the publication by Hahn and Meitner (in that order), in the *Physikalische Zeitschrift*, of "*Die Muttersubstanz des Actiniums*" ("The mother substance of actinium"). Actinium had been discovered by the friend and collaborator of Marie Curie, André-Louis Debierne in 1899. In their article, Hahn and Meitner propose to call the new element "protactinium" (Pa)[3].

Between 1920 and 1930, Lise's fame grew, she was recognized with several awards, and in 1926 she was named *nichtbeamteter ausserordentlicher professor* (a mouthful that meant she did not get paid, was not entitled to a pension, or to the privileges of a real professor in spite of "*ausserordentlicher*" meaning extraordinary) at the University of Berlin, the first in Germany, although this position was inferior to that of her peers. Everything changed when on January 30, 1933 Adolf Hitler assumed the position of *Reichskanzler*. It was the beginning of the Third Reich, the beginning of the end of German scientific primacy, and the beginning of a tragedy without equal: the ruthless extermination of European Jews and other minorities that would end twelve years later with the destruction of Germany and her allies, with an incalculable human and material cost, a testimony to human stupidity.

---

[3] Dmitri Ivanovich Mendeléyev had predicted in 1871 that there should be an element with atomic number 91 and that it would appear in the periodic table between thorium and uranium.

Figure 85 One among many. In the photo Lise is surrounded by the cream of physicists, in Berlin in 1920. From left to right: Otto Stern (Nobel Prize for Physics, 1943), Wilhelm Lentz, James Franck (Nobel Prize in Physics, 1925), Rudolf Ladenburg, Paul Knipping, Niels Bohr (Nobel Prize in Physics, 1922), E. Wagner, Otto von Baeyer, Otto Hahn (Nobel Prize in Chemistry, 1944), George de Hevesy (Nobel Prize in Chemistry, 1943), Lise Meitner, Wilhelm Westpahl, Hans Geiger, Gustav Hertz (with the pipe, Nobel Prize in Physics, 1925) and Peter Pringsheim. After Bohr's visit to Berlin, a great friendship resulted between Bohr and Meitner. Prof. Wilhelm Westfall, courtesy of AIP Emilio Segrè Visual Archives.

On April 7, 1933, the law for the Restoration of the Professional Civil Service was approved, excluding Jews from government service, which included university professorships. Although Jews represented only about one percent of the German population, they comprised twenty percent of the science faculties. They were expelled with little opposition from students and university colleagues. Max Planck appealed to Hitler to avoid the expulsions, to which he apparently replied: "If science cannot work without the Jews then we will have to function without science." In a letter written in May 1933 to Max Born (emigrated in 1933 to England), Einstein wrote: "I think you know, that I have never had a particularly favorable opinion of the Germans (speaking in political and moral terms). But I have to confess that the degree of brutality and cowardice has been something that has surprised me."

Figure 86 Three among many. The Physics Solvay conference of 1933. Seated from left to right: Schrodinger, Joliot-Curie, Bohr, Joffe, Curie, Langevin, Richardson, Rutherford, DeDonder, M. deBroglie, L. deBroglie, Meitner, Chadwick; Standing from left to right: Henriot, Perrin, Joliot, Heisenberg, Kramers, Stahel, Fermi, Walton, Dirac, Debye, Mott, Cabrera, Gamow, Bothe, Blackett, Rosenblum, Errera, Bauer, Pauli, Verschaffelt, Cosyns, Herzen, Cockcroft, Ellis, Peierls, Piccard, Lawrence, Rosenfeld. Note that Marie Curie and Irène did not sign but instead we read "Institut du Radium". Courtesy of the International Solvay Institutes, Brussels.

The result of this madness was the dismantling of German science and as one consequence the construction of the atomic bomb in the United States and the atomic bombing of Hiroshima and Nagasaki. Among the émigré scientists who contributed to that work were Hans Bethe, Felix Bloch, Albert Einstein, Enrico Fermi, James Franck, Otto Frisch, George Gamow, Emilio Segré, Leó Szilárd, Edward Teller, Stanislaw Ulam, Victor Weisskopf, and Eugene Wigner. Meitner would comment in a letter of 1946: "I cannot stop thinking about the kind of gift that Germany made to America."

Among the emigrants was Meitner's nephew, the physicist Otto Frisch (1904–1979), who left for England. By then she was acting as director of the KWI Physics Department, and although Jewish, the anti-Semitic laws did not directly apply to her, as she was Austrian. Nevertheless, she began to experience the tightening of freedom under the new authorities, including the revocation of her appointment as a professor at the University of Berlin. Little by little, she was not only losing space but also feeling the loss of her dear friends and colleagues who had emigrated. The absurd idea of an "Aryan physics", promoted by none other than the Nobel Prize winners Philipp Lenard (1862–1947) and Johannes Stark (1874–1957) (Nobel Prizes are no guarantee of decency) with the support of the Nazis, represented a clear threat to her livelihood and her research. When Austria was annexed, Lise's Austrian citizenship was no longer meaningful. And the dangers were getting closer and closer. Kurt Hess, a chemist at KWI, said bluntly: "The Jewish woman endangers our institute." It was time for Meitner to leave.

But she was not allowed to leave officially, although invited by several institutes in Holland, Sweden, and England (invitations extended with the purpose of getting her out), because, according to the authorities: "There are political objections to issuing a passport to Professor Meitner. It is considered undesirable for a recognized Jewess to leave Germany to act abroad, pretending to be a representative of German science, or with their names and corresponding experience to demonstrate their internal attitude towards Germany . . . The KWI will be able to find a way for her to work privately for the interests of the KWI after she resigns." Lise found herself "Damned if you do, damned if you don't."

Her friends, the Dutch physicists Dirk Coster (1889–1950) and Adriaan Fokker (1887–1972), aware of and worried about Lise's difficult circumstances, prepared a plan to help her escape. Under the pretext of job interviews, Coster traveled to Berlin in July of 1938. Lise was accompanied by Paul Rosbaud, editor of the prestigious *Naturwissenschaften* (and spy for the allies), to the train station with two suitcases, ten marks in her purse, and a diamond ring that Otto Hahn had inherited from his mother and given her. She boarded the train and, simulating a casual encounter, greeted Coster, who was already on his way back to Holland. They traveled with the knowledge that being discovered could have cost them their lives, and they managed to cross the border thanks to Coster and Fokker getting the collaboration of some Dutch border guards. Lise remembered the last moments of the trip[4]:

---

[4] George Axelsson (1946). Is the atom terror exaggerated? *Saturday Evening Post,* January 5.

At the Dutch border, I got the scare of my life when a Nazi military patrol of five men going through the coaches picked up my Austrian passport which had expired long ago. I was so frightened, my heart almost stopped beating. I knew that the Nazis had just declared open season on Jews, that the hunt was on. For ten minutes I sat there and waited, ten minutes that seemed like so many hours. Then one of the Nazi officials returned and handed me back the passport without a word.

Minutes later, she was safely across the Dutch border.

At fifty-nine, Meitner was about to begin a new and difficult phase of her long life. After the war she said: "Not only was it stupid but it was bad not to have left immediately"[5]. From Holland she traveled to Sweden, where she obtained a minor position at the Institute of Theoretical Physics in Stockholm, directed by Karl Manne Siegbahn (Nobel Prize in Physics, 1924), who did not like her. She felt lonely and deprived, she did not have her colleagues, her instruments, her books and papers, not even her personal belongings. She felt that she had been stripped of not only her belongings, but that which she most cherished—her physics. After a long process, the German authorities sent her some of her belongings, her chipped furniture, and her broken tableware.

Over the years while Lise's friends and colleagues had been fleeing Germany, scientific advances were occurring as well. In 1932, James Chadwick (1891–1974, Nobel Prize in 1935) discovered the neutron, and physicists began to use this particle for nuclear physics studies (having no electrical charge it can better penetrate the nucleus). Lise, aware of similar work done in the laboratory of Enrico Fermi (1901–1954, Nobel Prize in 1938) in Rome, was interested in studying reactions with uranium, and in 1934, she invited Hahn to join forces for these experiments. After a while he accepted and added the chemist Friedrich "Fritz" Strassmann (1902–1980) to the team. They worked together in a hostile environment, since the three identified themselves as opposed to the Nazi regime (and to top it off, one was Jewish). Strassmann had resigned his membership in the German Chemical Society when it became controlled by the Nazis; as he eloquently stated: "In spite of my affinity for chemistry, I value my personal freedom so much that to preserve it I would break stones to earn my living."

[5] John Cornwell (2004). *Hitler's scientists: science, war and the devil's pact*, pp. 207–13. Viking.

On December 19, 1938, Otto wrote a letter to Lise in Sweden. He told her about the results obtained in his laboratory together with Strassmann. By bombarding uranium with neutrons, they got something unexpected: barium atoms. He asked her if she had any idea how to explain these results. The answer: the uranium nuclei had been divided by the impact of the neutron into two parts, releasing a large amount of energy. Hahn published the experimental results, without mentioning Meitner, a fact that lends itself to different interpretations. Some understand that it had to do with the long shadow of Hitler, others interpret it as malicious. Under those circumstances, it was likely not possible for Hahn, in Berlin, to give credit to Meitner. In 1939, she published along with her nephew, the physicist Otto Frisch (exiled in England, son of her brother Auguste, composer and pianist), an article: "Disintegration of uranium by neutrons: a new type of nuclear reaction" in the prestigious journal *Nature*. They used the word "fission" for the first time, a name that Frisch gave to the process after an American biologist told him that the division of a bacterium was known as binary fission.

The idea of a chain reaction arose when in 1939 Enrico Fermi and Leó Szilárd (1898–1964) discovered that the fission of a uranium atom generated new neutrons, which in turn could induce new fissions. In December 1942, in Chicago, they were able to demonstrate this chain reaction experimentally for the first time. From there to the atomic bomb was a technological process that culminated in the first detonation on July 16, 1945. The Trinity Test, as it was called, took place in the desert of New Mexico, a spot appropriately called "Jornada del Muerto" or "Journey of the Dead" (a name given by the Spanish). Hahn received the 1944 Nobel Prize (not awarded until 1945) in Chemistry "for his discovery of the fission of heavy nuclei." Many protested, including Bohr, as they understood that Meitner and Frisch, and also Strassmann, should have been included. Although they were nominated in subsequent years they were not chosen[6]. It is also a fact that Hahn, having the authority to nominate others as a winner, did not nominate Meitner when he was able to do so in 1945 (he nominated Walther Bothe, who received it in 1954).

Dirk Coster, who helped her flee, wrote to Lise: "Otto Hahn, the Nobel Prize! Surely, he deserved it. But it is a pity that I took you out of

---

[6] Elisabeth Crawford, Ruth Lewin Sime, and Mark Walker (1997). A Nobel tale of postwar injustice. *Physics Today*, Vol. 50, No. 9, pp. 26–32.

Berlin in 1938 [. . .] Otherwise you too would be included, which would have been fair." Coster's take shows an almost charmingly naïve neglect of the political context; Renate Feyl sums it up more succinctly[7]: "Her work was crowned with the Nobel Prize for Otto Hahn."

Lise was under no misapprehensions about the political significance of her—and Hahn's—decisions. A mere six weeks after the surrender of Germany, Lise wrote a letter to Otto on June 27, 1945 that said in part:

You all worked for Nazi Germany and you did not even try passive resistance. Granted, to absolve your consciences you helped some oppressed person here and there, but millions of innocent people were murdered and there was no protest. I must write this to you, as so much depends upon your understanding of what you have permitted to take place. Here in neutral Sweden, long before the end of the war, there was discussion of what should be done with German scholars when the war is over. What then must the English and the Americans be thinking? I and many others believe one path for you would be to deliver an open statement that you are aware that through your passivity you share responsibility for what has happened, and that you have the need to work for whatever can be done to make amends. But many think it is too late for that. These people say that you first betrayed your friends, then your men and your children in that you let them give their lives in a criminal war, and finally you betrayed Germany itself, because even when the war was completely hopeless, you never once spoke out against the meaningless destruction of Germany. That sounds pitiless, but nevertheless I believe that the reason that I write this to you is true friendship. You really cannot expect that the rest of the world feels sympathy for Germany. In the last few days one has heard of the unbelievably gruesome things in the concentration camps; it overwhelms everything one previously feared. When I heard on English radio a very detailed report by the English and Americans about Belsen and Buchenwald, I began to cry out loud and lay awake all night. And if you had seen those people who were brought here from the camps. One should take a man like Heisenberg and millions like him, and force them to look at these camps and the martyred people. The way he turned up in Denmark in 1941 is unforgettable.

Perhaps you will remember that while I was still in Germany (and now I know that it was not only stupid but very wrong that I did not leave at once) I often said to you: as long as only we have the sleepless nights and not you, things will not get better in Germany. But you had no sleepless nights, you did not want to see, it was too uncomfortable. I

---

[7] Renate Feyl (1994). *Der lautlose Aufbruch, Frauen in der Wissenschaft*, p. 199. Köln.

could give you many large and small examples. I beg you to believe me that everything I write here is an attempt to help you. [The letter, given to a military messenger, was never delivered to Hahn.]

Despite her suffering and forced emigration, despite being excluded from Hahn's Nobel Prize (which in her words was "in no way an open wound"[8]), despite Hahn publicly highlighting his own work and belittling Meitner's contribution (and leadership) in the investigations with uranium (a posture that he maintained during all of his life and that did, in fact, bother Lise), she preserved her links with him and was magnanimous. However, although she maintained her friendship with Hahn (in their letters she calls him "Hähnchen"), with whom she had collaborated for thirty years, it was no longer the same. Meitner not was troubled by exclusion from awards; what did pain her was to be known as "Hahn's collaborator", and, more broadly, that her scientific colleagues tried to sweep the difficult past under the rug, instead of examining their behaviors and facing the harsh reality of the world and their decisions in it.

Her exile in Sweden, where she did not have a laboratory nor support staff, like she did in the KWI, limited her in what she could accomplish scientifically, and this more than other limitations caused her distress. These circumstances also prevented her from discovering the first transuranic element (number 93), although she knew how to find it and made an attempt to isolate it on a visit to the Bohr institute in Copenhagen, just as the Germans were invading Denmark (putting an end to this series of experiments). The discovery was made seven weeks later in Berkeley, California, by Edwin McMillan (1907–1991) and Phillip Ableson (1913–2004), who called it "neptunium" (since it follows uranium, as the planet Neptune follows Uranus). McMillan received the 1951 Nobel Prize in Chemistry together with Glenn Seaborg (1912–1999).

In January 1946, at the age of sixty-eight, Lise Meitner left her exile having been invited as a visiting professor to the Catholic University of America in Washington DC. She was received as she deserved, overwhelmed by the press ("the mother of the atomic bomb"), and celebrated by many. The Woman's National Press Club designated her "woman of the year" and she shared the podium with President Harry Truman. She traveled to several universities to give lectures and meet with old acquaintances in physics and to make new ones, including

[8]   Ruth Lewin Sime (1996). *Lise Meitner: a life in physics*, p. 374. University of California Press.

Einstein, Weyl, Fermi, Teller, and Szilárd. She returned to her world, the world of physics, and she felt happy.

Upon the expiration of her six-month-long visiting professorship at Catholic University, she returned via London, to an improved situation in Sweden, with new laboratories at the Royal Institute of Technology, following a great interest on the part of the Swedish government to start a nuclear physics program. She was apprehensive about meeting Hahn, who would be in Stockholm to receive the Nobel Prize (awarded to him in 1944).

Her stance in relation to the behavior of her German colleagues during Nazism, and especially in the postwar period, where the excuse of state terror could no longer be wielded (as she expressed in her letter to Hahn), prevented her from returning permanently to Germany (something that happened with many exiles—Einstein never stepped on German soil again), although she was invited to direct a new institute of chemistry by Strassmann. In a letter she wrote[9]: "I think I could not live in Germany. After everything I see in the letters of my German friends, and what I hear about Germany, I conclude that the Germans have not yet understood what happened, and all the atrocities that do not affect them personally are completely forgotten." She insisted that the Germans should recognize their lack of action against Nazism, and that it was not acceptable that they should now be seen as "victims." In the world of physics in particular, she thought little of the fabrication of Weizsäcker and Heisenberg, who claimed that the Germans had been capable of manufacturing the bomb, but they did not do it for ethical reasons (according to the physicist Goudsmit: "A brilliant rationalization of failure"), whereas the vision from the outside was that they were simply not able (because they had driven out the talent who could do it!).

In April of 1948, she visited Germany for the first time (after ten years), invited to participate in a ceremony in Göttingen in memory of her dear friend Max Planck, recently deceased. During her visit, walking through the well-known streets, she could not avoid looking at people and thinking what they would have been doing in the last ten years.

With the passage of time, she was able to overcome some of her reservations, and take part in conferences and meetings in Germany, although always with an element of distrust in her relations with her

---

[9] Charlotte Kerner (1986). *Lise, Atomphysikerin. Die Lebensgeschichte der Lise Meitner*. Beltz.

old colleagues, including Hahn, who was never able to recognize the importance of Meitner in the discovery of nuclear fission.

For many years, the idea spread (backed by Hahn) that she had nothing to do with the discovery of nuclear fission. It is an example of the "Matthew effect", so baptized by the sociologist Robert K. Merton, describing how, as in many other arenas of life, eminent scientists will often get more credit than a comparatively unknown researcher, even if their work is similar; it also means that credit will usually be given to researchers who are already famous. (The name refers to Matthew 25:29: "For to everyone who has will more be given, and he will have abundance; but from him who has not, even what he has will be taken away."). For example, a prize will almost always be awarded to the most senior researcher involved in a project, even if all the work was done by a graduate student (as was the case for the Nobel Prize in Physics for 1974 awarded to Anthony Hewish for the discovery of pulsars, work which was done under his supervision by Jocelyn Bell Burnell (born 1943). In 1949, Meitner and Hahn received the Max Planck medal from the German Society of Physics.

After the Swedish nuclear project culminated with the commissioning of the first Swedish nuclear reactor, she retired in 1960, at the age of eighty-one, and moved to Cambridge, England, to be near her beloved nephew Otto Frisch and his family.

In 1966, as if to rectify history, the Enrico Fermi Prize was awarded to Hahn, Meitner, and Strassmann "for their collaborative and independent contributions to the discovery of nuclear fission." Lise and Otto, both eighty-eight years old, were unable to attend the ceremony in Washington; Hahn and Strassmann received the award in Vienna, and Meitner in Cambridge, from Glenn Seaborg (director of the United States Atomic Energy Commission) and Otto Frisch. She died in 1968, shortly after the death of her controversial friend Otto Hahn. The epitaph on her grave was composed by her nephew Frisch: "Lise Meitner: A physicist who never lost her humanity."

We observe a certain parallelism between the pairs of Meitner and Hahn and Skłodowska and Curie, in the sense that they worked together effectively, they devoted themselves to the study of the phenomenon of radioactivity, they discovered new elements, and the quality and novelty of their research led to Nobel Prizes. But there is also orthogonality. Marie and Pierre loved each other, and Pierre protested Marie's exclusion, successfully including her in the recognition. Lise (eleven years younger than Marie) and Otto were friends, but Otto took her out of his professional life when it was convenient and did not want to acknowledge her great work or nominate her (when he could have done so) for a well-deserved Nobel Prize. Both women were touched by tragedy: Marie with the death of her beloved Pierre in 1906, after eleven years of happy marriage, and Lise with expulsion from Germany in 1938, after a lifetime dedicated to research in the KWI, much of it with Otto.

In 1982, at the Institute for Heavy Ion Research (*Gesellschaft für Schwerionenforschung*), near Darmstadt ("city of the intestine"), the element with atomic number 109 was synthesized, by bombarding atoms of bismuth with iron nuclei. It is a radioactive element that does not exist naturally and which has a half-life of only 7.6 seconds. In 1997, the international commission dedicated to these matters (International Union of Pure and Applied Chemistry) agreed that this element would be named "meitnerium." The asteroid 6999 Meitner, discovered on October 16, 1977, bears her name as well. And in 1970, the IAU dedicated a crater of about 90 km on the far side of the Moon to her, not very distant from Sklodowska.

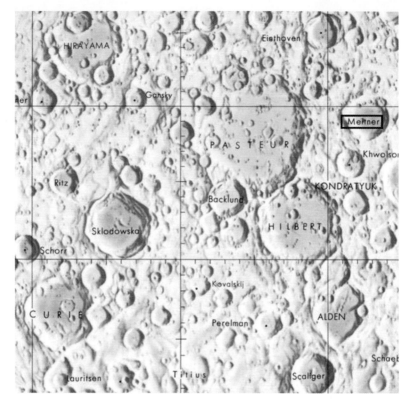

Figure 87 Location of crater Meitner. Courtesy of the Lunar and Planetary Institute, Houston, Texas.

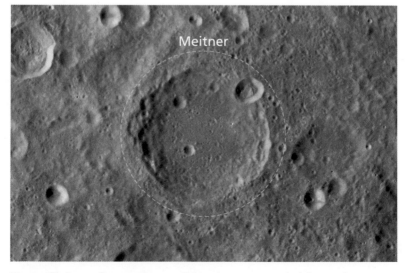

Figure 88 Lunar Reconnaissance Orbiter zoom on crater Meitner (image width is 150 miles).

20

Amalie Emmy Noether (1882–1935)

Figure 89 Emmy Noether. Courtesy of Drs. Emiliana and Monica Noether.

If one proves the equality of two numbers a and b by showing first that "a is less than or equal to b" and then "a is greater than or equal to b", it is unfair, one should instead show that they are really equal by disclosing the inner ground for their equality.

EMMY NOETHER

We are of the opinion that a female head can engage in creative research in mathematics only in a highly exceptional case.

From a letter of 1915 addressed to the Minister of Education and Intellectual Affairs by the Division of Mathematics and Natural Sciences of the Faculty of Philosophy of the University of Göttingen[1]

The mathematician Emmy Noether was born thirty-two years after Sofia Kovalévskaya, but she faced not only difficulties because of her gender, but also those caused by the beasts of Nazism, because, like Meitner, she was Jewish.

In the field of theoretical physics, she discovered a beautiful and fundamental theorem in 1915, published and presented by the Göttingen

---

[1] Cordula Tollmien (1990). "Sind wir doch der Meinung, daß ein weiblicher Kopf nur ganz ausnahmsweise in der Mathematik schöpferisch tätig sein kann..."—eine Biographie der Mathematikerin Emmy Noether (1882–1935) und zugleich ein Beitrag zur Geschichte der Habilitation von Frauen an der Universität Göttingen. *Göttinger Jahrbuch*, Vol. 38, pp. 153–219.

The Women of the Moon. Daniel R. Altschuler Stern and Fernando J. Ballesteros Roselló.
© Daniel R. Altschuler Stern and Fernando J. Ballesteros Roselló 2019. Published in 2019 by Oxford University Press. DOI: 10.1093/oso/9780198844419.001.0001

mathematician Felix Klein in 1918, which connects two fundamental pillars of physics: symmetries in nature and conservation laws.[2]

A symmetry in nature occurs when it is possible to make a change in some circumstance of an experiment, so that after the change the same results are obtained. For example, you can move a laboratory to another planet and verify that the laws of physics in the new location have not changed, which illustrates a symmetry of nature under translation. In the same way, the laws of physics are invariant to a translation in time (that is, the result of an experiment does not change if we do it today or next year). We can see, by observing distant stars (another site in another time), phenomena that obey the same laws of quantum mechanics that we observe in terrestrial laboratories. If it were not so, we would not be able to understand anything.

These symmetries correspond to important conservation laws: there is a quantity that can be measured before and after some process, and that quantity is conserved in the process, that is, it has the same value before and after. In the case of symmetry over time, Noether showed that it corresponds to energy conservation. Similarly, she proved that from the fact that physics is symmetric with respect to a translation in space, the law of conservation of momentum (mass times velocity) follows. Each symmetry corresponds to a conservation law and each conserved quantity points to a symmetry. These are results that are fundamental in modern physical theories, and are considered as important as those of Einstein. Noether, in short, tells us that the laws of nature arise from the geometry of space-time. However, Emmy is a relatively unknown character even in professional circles, aptly described by Natalie Angier as[3] "The mighty mathematician you've never heard of".

Emmy was born in 1882, daughter of the well-known mathematician Max Noether (1844–1921), professor at the University of Erlangen and one of the founders of algebraic geometry, and Ida Amalia Kaufmann (1852–1915), pianist and daughter of a prosperous merchant. She followed the typical educational path of those times in a school for women

[2] Klein presented it for her, since a woman could not belong to the *Königliche Gesellschaft der Wissenschaften zu Göttingen* (Royal Society for the Sciences of Göttingen). The Royal Society of London, founded in 1662, elected its first female members in 1945 (Marjory Stephenson, biochemist, and Kathleen Lonsdale, physicist); and the Académie des Sciences, in Paris, founded in 1666, in 1962 (Marguerite Catherine Perey, physicist).

[3] Natalie Angier (2012). The mighty mathematician you've never heard of. *The New York Times*, Science section, March 26.

(*Höhere Töchter Schule*), which concluded in 1900 with a certification for teaching English and French. But to the surprise of many, she decided that she wanted to study mathematics. The University of Erlangen, where her older brother Fritz was studying, did not allow her to enroll, but allowed her to attend courses as auditor (with prior permission of the professor), one of two women among thousands of students. After two years of study she took the entrance exams to the universities in Bavaria and passed them. But she decided to go to the University of Göttingen, in Prussia, in order to take classes (again, as auditor) with luminaries of the era: David Hilbert (1862–1943), Felix Klein (1849–1925), and Hermann Minkowski (1864–1909). A year later, when Erlangen began admitting women, she returned and matriculated.

After five years of study and a thesis under the supervision of Professor Paul Gordan (1837–1912), a mathematician friend of the family, she obtained her doctorate *summa cum laude* in 1907.

Nevertheless, it was not easy for her to get a job. The German universities did not allow women on their faculties and Emmy decided to help her father, crippled by polio, in the Mathematics Institute at Erlangen, substitute lecturing for him in some classes, while preparing her first publications.

Consequently, she worked without pay for seven years. But gradually she was shaping up as a great mathematician, and began to be recognized as such. At first, she was known as Max Noether's daughter, today Max is known as Emmy Noether's father. In 1908 she was accepted as a member of the *Circolo Matematico di Palermo*, and in 1909 she was elected as a member of the *Deutsche Mathematiker-Vereinigung* and gave her first public presentation at the 1909 annual meeting in Salzburg. (She was never elected to the *Göttingen Gesellschaft der Wissenschaften*, an honor conferred on several of her colleagues.)

Following the death of Gordan, Max and his daughter visited David Hilbert and Felix Klein in Göttingen, who recognized Emmy's genius. In 1915, she was invited by Hilbert and Klein to the University of Göttingen, at that time considered the world's capital of mathematics. The faculty was opposed and one of the professors said: "What will our soldiers think when they return to the university and find that they are required to learn at the feet of a woman?" Hilbert, outraged by this attitude, replied[4]: "I do not see that the sex of the candidate is an argument

[4] Sharon Bertsch McGrayne (2001). *Nobel Prize women in science: their lives, struggles, and momentous discoveries*, p. 73. Joseph Henry Press.

against her, after all, we are a university and not a bathhouse." Despite the lack of formal support, Emmy moved to Göttingen in 1916, and lived on the money her family sent her.

After the defeat of Germany, the German Republic was established. Women were granted the right to vote, a step toward full emancipation. Towards 1919, in a Germany convulsed by the ravages of war under a socialist government, and with the support of Einstein, Hilbert and Klein, the Prussian Ministry of Education allowed, after much tug-of-war, Noether to be named Privatdozent, the lowest academic rank, on the condition that she give classes as an assistant to Hilbert. This was done; a mathematical physics class was announced, but Noether lectured: "Seminar on Mathematical Physics: Professor Hilbert, with the assistance of Dr E Noether, Monday from 4-6."

In 1921 she published a monumental paper that laid the foundations of many future studies, and is considered one of her best[5]: "*Idealtheorie in Ringbereichen.*"

Finally, in 1922, Emmy was appointed to the position of *Nichtbeamteter Ausserordentlicher Professor* and was able to give her own classes. Discrimination did nothing to slow down her work, since she had very little interest in material things. Her research and her classes captivated a small group of fans who admired her, and they became known as the "Noether boys". She supervised a total of sixteen doctoral theses, two by women.

She is described as a happy woman, completely dedicated to mathematics, who did not care about her appearance, and who was frank, friendly, generous, and overweight—a larger-than-life figure, especially in contrast to the expectations of a woman of her class and time. The Czech mathematician Olga Taussky (1906–1995) relates[6] that at the dinner of a meeting of the *Deutsche Mathematiker-Vereinigung* in 1930: "Emmy was very entertained discussing mathematics with the people around her. She was having a great time. While eating, she gesticulated with enthusiasm as the food fell on her dress. Emmy simply wiped it from her dress, undeterred and continued the discussion". (In 1971, Taussky became the first female professor at the California Institute of Technology, Caltech.)

[5] Emmy Noether (1921). Idealtheorie in Ringbereichen. *Mathematische Annalen*, No. 83, pp. 24–66.
[6] Olga Taussky (1981). My personal recollections of Emmy Noether. In James W. Brewer and Martha K. Smith (eds), *Emmy Noether: a tribute to her life and work*, pp. 79–92. Marcel Dekker.

It was during the years in Göttingen when she developed her most important work, including her theorem on symmetry and conservation, which we have already presented. In the 1920s, she also published important results in abstract algebra, collaborated with European mathematicians, trained students, and held the position of editor of the *Mathematische Annalen*, the most prestigious professional journal of the time. In the winter of 1928–1929 she accepted an invitation from the renowned mathematician Pável Serguéyevich Aleksándrov (1896–1982), to visit Moscow State University, where she collaborated with Russian mathematicians, and in 1932 she received the Alfred Ackermann-Teubner-Gedächtnispreis

Figure 90 Photo taken in July 1933. From left to right are Ernst Witt (who had joined the Nazi party and the Sturmabteilung as of May 1933), Paul Beranys, Helene Weyl, Hermann Weyl (about to leave Nazi Germany with his whole family to become a permanent member of the newly created Institute for Advanced Study), his son Joachim, Emil Artn, Emmy, Ernst Knauff, unknown, Zeng Jiongzhi (seated on the right, one of her students from China) and Ema Bannow. For interesting details see: *Dating the Gasthof Vollbrecht Photograph* by Christophe Eckes and Norbert Schappacher, January 5, 2016. Photo provided by Artin, Natascha. Oberwolfach Photo Collection.

zur Förderung der Mathematischen Wissenschaften, a prestigious recognition.

With the rise of Hitler to power in 1933, that close to ideal world began to crumble. Emmy—Jewish, intellectual, pacifist, and socialist—was one of the first to be fired from the university by the Nazis, along with other eminent Jewish colleagues, some of them deported to concentration camps and to their deaths. Similar tragedies occurred throughout Germany; emigration, when possible, was still the difficult and dangerous alternative to death.

Emmy continued for a year tending to her "boys" in her small apartment, not flinching when someone appeared wearing the brown shirt of the Nazi party adorned with a swastika. But the situation was unsustainable, and the risk high.

Shortly after this debacle, during a banquet, the new Nazi Minister of Education (a nice oxymoron), Bernhard Rust (1883–1945, by suicide), asked Hilbert if it was true that the Institute of Mathematics had suffered with the dismissal of the Jews; Hilbert replied: "The institute? It really does not exist any longer."

Both Einstein (1878–1955, Jewish) and Hermann Weyl (1885–1955, married to a Jew) emigrated to the United States in 1933, and there they became part of the Institute for Advanced Studies of Princeton, where, together with other illustrious immigrants such as Kurt Gödel and John von Neumann, they formed a group of the highest intellectual caliber.

With the help of Einstein and Weyl, and the support of the Rockefeller Foundation, Emmy emigrated in 1934 to take a one-year post at Bryn Mawr College, a women's college in Pennsylvania, not far from Princeton, and Einstein. At Bryn Mawr, she became friends with the mathematician Anna Pell Wheeler (1883–1966), who had invited her and had studied at Göttingen. She was well received by her students, who for the first time were being taught by the mathematician who had more or less singlehandedly created the subject of the course. She again started with a small group: this time they were the "Noether girls".

Weyl, referring to the Nazi regime, said: "A stormy time of struggle like this one we spent in Göttingen in the summer of 1933 draws people closer together; thus, I have a particularly vivid recollection of these months. Emmy Noether, her courage, her frankness, her unconcern about her own fate, her conciliatory spirit, were in the midst of all the hatred and meanness, despair and sorrow surrounding us, a moral solace."

After a year at Bryn Mawr she returned to Germany to collect her belongings, and to see (for the last time, it would turn out) her mathematician brother Fritz (1884–1941), also fired, who emigrated with his wife and two children to a position at the University of Tomsk, in Siberia. During her second year at Bryn Mawr, Ruth Stauffer (1910–1993), her first student, received her doctorate. Noether traveled weekly during this year to the Institute of Advanced Studies at Princeton, to give a seminar on abstract algebra to some "students," none other than Einstein, Alfred T. Brauer, Abraham Flexner (1866–1959, founder and director of the Institute for Advanced Studies), Solomon Lefschetz, Oswald Veblen (1880–1960, North American mathematician), and Hermann Weyl.

Some of her colleagues in the scientific community were concerned because they thought that Noether was wasting her time teaching undergraduate students and that she deserved a position at a university that had top-notch graduate programs to continue her research work. A plan was drawn up so that eventually she would join the Institute for Advanced Studies.

Sadly, this would not come to pass. As a result of an operation to remove a tumor in her ovary, she contracted an infection that caused her death at fifty-three years of age. The community of physicists and mathematicians mourned the loss.

Two years later, in 1937, her brother Fritz was arrested by the Soviet authorities during the "great purge," accused of espionage, and in 1941 was shot for "anti-Soviet propaganda." The efforts of Einstein and Weyl failed to rescue him, but resulted in his two sons, Gottfried (1915–1991) (who was also a mathematician) and Herman (1912–2007) (who was a chemist), migrating to the United States.

Albert Einstein wrote in a letter to *The New York Times* on the occasion of Emmy's death[7]:

Within the past few days a distinguished mathematician, Professor Emmy Noether, formerly connected with the University of Göttingen and for the past two years at Bryn Mawr College, died in her fifty-third year. In the judgment of the most competent living mathematicians, Fräulein

[7] Professor Einstein Writes in Appreciation of a Fellow-Mathematician. *The New York Times*, May 5, 1935.

Noether was the most significant creative mathematical genius thus far produced since the higher education of women began. In the realm of algebra, in which the most gifted mathematicians have been busy for centuries, she discovered methods which have proved of enormous importance in the development of the present-day younger generation of mathematicians. Pure mathematics is, in its way, the poetry of logical ideas. One seeks the most general ideas of operation which will bring together in simple, logical and unified form the largest possible circle of formal relationships. In this effort toward logical beauty spiritual formulas are discovered necessary for the deeper penetration into the laws of nature.

Hermann Weyl[8], in a ceremony in memory of Emmy, said:

I earnestly tried to obtain from the Ministerium a better position for her because I was ashamed to occupy such a preferred position beside her whom I knew to be my superior as mathematician in many respects. I did not succeed, nor did an attempt to push through her election as a member of the Göttingen Gesellschaft der Wissenschaften. Tradition, prejudice, eternal considerations, weighted the balance against her scientific merits and scientific greatness, by that time denied by no one. In my Göttingen years, 1930–1933, she was without doubt the strongest center of mathematical activity there, considering both the fertility of her scientific research program and her influence upon a large circle of pupils [. . .]

For two of the most significant sides of the general relativity theory she gave at that time the genuine and universal mathematical formulation.

Later, Weyl compares her to another of the women of the Moon, Sofia (Sonya) Kovalévskaya:

Sonya had certainly the more complete personality, but was also of a much less happy nature. In order to pursue her studies Sonya had to defy the opposition of her parents, and entered into a marriage in name only, although it did not quite remain so. Emmy Noether had, as I have already indicated, neither a rebellious nature nor Bohemian leaning. Sonya possessed feminine charm, instincts and vanity; social successes were by no means immaterial to her. She was a creature of tension and whimsy; mathematics made her unhappy, whereas Emmy found the greatest

---

[8] H. Weyl, published in *Scripta Mathematica*, III. 3, pp. 201–220, 1935.

pleasure in her work. [. . .] But Emmy Noether without doubt possessed by far the greater power, the greater scientific talent.

Leon Lederman, Nobel Prize in Physics (1988), says about Noether[9]:

Emmy Noether made one of the most significant contributions to human knowledge through her remarkable theorem. The theorem cleanly and clearly unites symmetries with the complex dynamics of physics and forms a basis for human thought to make forays into the inner world of matter at the most extreme of energies and distances. One might argue that Noether's theorem is as important to under-standing the dynamic laws of nature as is the Pythagorean theorem to understanding geometry.

Emmy has an asteroid to her name: 7001 Noether, discovered in 1955. And in 1970, the IAU named a 67 km diameter crater in her honor located on the far side of the Moon.

[9] Leon Lederman and Christopher Hill (2004). *Symmetry and the beautiful universe.* Prometheus.

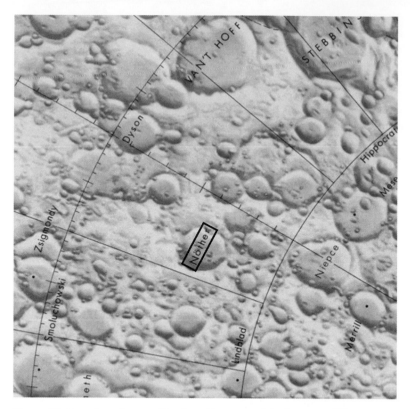

Figure 91 Location of crater Noether. Courtesy of the Lunar and Planetary Institute, Houston, Texas.

Figure 92 Lunar Reconnaissance Orbiter zoom on crater Noether (image width is 150 miles).

21

Louise Freeland Jenkins (1888–1970)

Figure 93 Louise Freeland Jenkins. Mount Holyoke College Archives and Special Collections.

Two special pleasures of my first year here [in the residence] were to see the sometimes-wonderful coloring in the sky just before sunrise, while I am still in bed. The other was the coloring of the autumn when the leaves turned such varied coloring when the nights became colder, some stayed green, then nearby would be yellow, red, orange, brown, etc. with each a different color all down the line, at the border of our land. And each day the colors were different from the day before. The beauty of God's world, and His loving care of us is precious to me.

From a letter Miss Jenkins wrote in 1970 to Mrs. Sato Natasaki, her student at the Women's Christian College of Japan, and a fellow teacher at the Hinomoto Gakuen girls' high school[1]

Louise Freeland Jenkins may be the most enigmatic of the women of the moon. With the exception of Hypatia—about whom we might select from abundant myth—her personal life is largely a black box. Her professional work, however, survives; she greatly advanced our three-dimensional map of the universe over the course of the twentieth century.

She was born on July 5, 1888, in Fitchburg, Massachusetts. We know nothing of her family and her youth, and she does not appear among

[1] Sei-ichi Sakuma (1985). Louise F. Jenkins, astronomer and missionary in Japan. *Journal of the American Association of Variable Star Observers*, Vol. 14, No. 2. http://www.aavso.org/media/jaavso/819.pdf

The Women of the Moon. Daniel R. Altschuler Stern and Fernando J. Ballesteros Roselló.
© Daniel R. Altschuler Stern and Fernando J. Ballesteros Roselló 2019. Published in 2019 by Oxford University Press. DOI: 10.1093/oso/9780198844419.001.0001

the eighteen hundred biographies contained in the *Notable American Women, 1607–1950*.[2]

In 1911, she graduated from Mount Holyoke College[3] with a bachelor's degree, and earned her master's degree in astronomy there in 1917. She set immediately to her life's work of astronomy and education.

Louise specialized in the observation of variable stars and the determination of trigonometric parallaxes of stars to calculate their distances. She worked as an astronomy assistant at Mount Holyoke (1911–1913), as a computer at the Allegheny Observatory in Pittsburgh (1913–1915), under the direction of astronomer Frank Schlesinger (1871–1943), and then as an instructor at Mount Holyoke (1915–1920). Jenkins observed sunspots with the Mount Holyoke telescope, reporting her observations in *Popular Astronomy*, and in 1919 she became a member of the American Association of Variable Star Observers (AAVSO).

In 1920, she joined the Woman's American Baptist Home Mission Society, and traveled to Japan as a missionary. Before leaving, another member of the AAVSO, Charles Elmer (1872–1954), gave her a three-inch telescope to take to Japan. He was one of the founders of the Perkin Elmer Corporation, today a multinational corporation with over 10,000 employees dedicated to the production of scientific and medical instruments.

Perkin Elmer's connection with astronomy has continued over the decades. Between 1979 and 1981 this company was in charge of the construction of the main mirror of the Hubble Space Telescope. Once in space (carried by Space Shuttle Discovery in 1990), and to the dismay of those in charge at NASA, it was found that a measurement error had caused an error in the curvature of the mirror. It was a tiny error, just a twenty-fifth of the width of a human hair. The microdefect was enough to make the telescope, which cost 1600 million dollars, suffer from myopia, giving us images never before seen of an extraordinary universe . . . blurry. Surprisingly, no check had been made before the launch. If that weren't frustrating enough: there was a reserve mirror on the ground that was perfectly formed. This led to a bitter discord

---

[2] Edward T. James, Janet Wilson James, and Paul Boyer (eds.) (1950). *Notable American Women, 1607–1950: a biographical dictionary*. Belknap Press of Harvard University Press.

[3] Mount Holyoke College, founded in 1837 as the Mount Holyoke Female Seminary, is an exclusive university (the current cost per year over $65,000 including room and board) for women (even in the present, but also accepting transgender students), located in South Hadley, Massachusetts.

between Perkin Elmer and NASA. It was necessary to build a corrective lens for the telescope, which was installed in orbit by the crew of the Space Shuttle Endeavor in 1993. The Hubble has since performed wonders.

Let's return to Jenkins. In Tokyo she taught astronomy at the Women's Christian College, along with English and Bible studies. From there she also observed variable stars and reported her observations to the AAVSO. Unfortunately, the great earthquake of September 1, 1923, a brutal event that devastated Tokyo and cost the lives of 150,000 people, destroyed her instrument, although fortunately she was in Karuizawa at that time, 100 km from Tokyo.

In 1925, she came home to Massachusetts due to the death of her father, returning to Japan in 1926 as a teacher and dormitory superintendent at the Hinomoto Gakuen school for women near Kyoto.

In 1932, she returned to the United States to work at the Yale Astronomical Observatory, whose director at the time was Frank Schlesinger (known as the father of modern astrometry), with whom Jenkins had worked at the Allegheny Observatory some twenty years earlier. Schlesinger, like Pickering, held dimly progressive ideas about the role of women common to astronomy at that time. In a letter written in 1901 to the astronomer George Ellery Hale (1868–1938) we can read[4]: "I am thoroughly in favor of employing women as measurers and computers and I think their services might well be extended to other departments. Not only are women available at smaller salaries than are men, but for routine work they have important advantages. Men are more likely to grow impatient after the novelty of the work has worn off and would be harder to retain for that reason."

Between 1942 and 1958, Jenkins was assistant editor of the important *Astronomical Journal*. She continued to work on the determination of stellar parallaxes and together with Schlesinger published the second edition of the Yale catalog of bright stars in 1940. Through 1962 she had determined (or supervised the determination) of 350 parallaxes, an arduous and difficult task, measuring the tiny angular displacement of a star when observed from different points of the Earth's orbit around the Sun. To give a sense of scale, Proxima Centauri, the closest star to

[4] Frank Schlesinger to George Ellery Hale, July 1, 1901. Quoted on p. 370 in E. Dorrit Hoffleit (2002). Pioneering women in the spectral classification of stars. *Physics in Perspective*, Vol. 4, pp. 370–98.

the Sun (and therefore, with the largest parallax), is about four light-years away and its parallax is only 0.77 seconds of arc (one second of arc is 1/3600 of one degree). With time and better instrumentation, parallaxes up to about 0.01 seconds of arc have been measured corresponding to distances of about 300 light-years for a few stars (of course, with lower precision for larger distances).

We can evaluate how astronomical techniques have advanced by comparing these distances with the measurement of the Hipparcos satellite (High Precision Parallax Collecting Satellite), of the European Space Agency (ESA), launched in 1989, which managed to measure the distance of some 100,000 stars with an error smaller than 10 percent reaching 500 light years. The ESA's GAIA[5] (Global Astrometric Interferometer for Astrophysics) mission, which was put into orbit around the Sun in December 2013, has measured the position and distance of approximately 1.3 billion stars, a huge amount (some would say an astronomical amount) that, however, only represents one percent of the stars in our Galaxy. Its angular resolution is equivalent to observing a United States quarter or a one-euro coin on the surface of the Moon. GAIA can measure the distance of a star to 30,000 light years with an uncertainty of twenty percent.

In September 1957, Jenkins returned to Japan and visited the Tokyo Observatory as a professional astronomer, just one day before the launch of Sputnik. In her visit to the International Christian University, she fell and fractured her leg, returning in a wheelchair to the United States after a month's stay in hospital. She continued with her task of determining parallaxes until 1968. She died in 1970 in a nursing home in New Haven, Connecticut.

Like the parallaxes Jenkins herself observed and measured, the crater Jenkins is visible from Earth, but requires immensely precise observation at particular points in the orbital cycle. It is located at the very eastern end of the visible face, right on the edge of the limb.

---

[5] http://sci.esa.int/gaia/

Figure 94 Location of crater Jenkins. Courtesy of the Lunar and Planetary Institute, Houston, Texas.

Figure 95 Lunar Reconnaissance Orbiter zoom on crater Jenkins (image width is 150 miles).

Figure 96 Some attendees of the 17th meeting of the Astronomical and Astrophysical Society of America visiting Yerkes Observatory. The meeting was held at Dearborn Observatory, Northwestern University, Evanston, Illinois, August 25–28, 1914, at which time the group officially adopted a new name, the American Astronomical Society Annie J. Cannon is the fourth on the top row (dressed in white). On the second row from the bottom we see Edwin P. Hubble and to his left sits Louise F. Jenkins. The lady with the flowery hat is Sarah F. Whiting (Annie's professor at Vassar). To the left among those standing, the one in the middle is Edward C. Pickering (president of the society). University of Chicago Photographic Archive [apf6-00393]. Special Collections Research Center, University of Chicago Library.

22

Priscilla Fairfield Bok
(1896–1975)

Figure 97 Priscilla Bok (photo taken in 1969). AIP Emilio Segre Visual Archives, John Irwin Slide Collection.

If I have to present myself soon to Saint Peter, I think I'll ask him to give me a front row seat in the center of the nebula. So I can see the stars forming right in front of my eyes.

PRISCILLA to her
husband (1975)

Priscilla Fairfield was born in Littleton, Massachusetts, in 1896, and was another of the "Women of the Moon" who was linked to the Harvard Observatory, although she belonged to a later generation than "Harvard's computers." Very little has come to us from her childhood, except that from a very young age she felt a passion for astronomy and that her family was not wealthy, so Priscilla had to work to pay her own tuition at Boston University.

She quickly showed herself to be a precocious researcher, and as an undergraduate, she followed two years of sunspots with the solar telescope of the Judson B. Coit Observatory, on the roof of the Faculty of Arts and Sciences of Boston University. Priscilla had to bribe the school guard to allow her access to the telescope during weekends[1]. The result of this

---

[1] David H. Levy (1993). *The man who sold the Milky Way: a biography of Bart Bok*. University of Arizona Press.

The Women of the Moon. Daniel R. Altschuler Stern and Fernando J. Ballesteros Roselló.
© Daniel R. Altschuler Stern and Fernando J. Ballesteros Roselló 2019. Published in 2019
by Oxford University Press. DOI: 10.1093/oso/9780198844419.001.0001

research, published before finishing her bachelor's degree, was her first scientific article, *Observations of sunspots at Boston University*, and the first in a long list.

After graduating in 1917, she moved to the University of California at Los Angeles (UCLA) to work towards a doctorate at UCLA's Lick Observatory. There she learned spectroscopy with her thesis supervisor, William Wallace Campbell (1862–1938). She was his last PhD student, in fact, because after the direction of Priscilla's work, Campbell became the tenth President of the University of California system from 1923 to 1930. Perhaps this coincidence played a part in the direction of their relationship, but the warmth and calmness of Priscilla would likely have been sufficient for him to feel great affection for her—they remained good friends for the rest of his life.

After a mere three years of work at the Lick Observatory, Priscilla attained her doctorate in astronomy in 1921. A year later, she was hired by Smith College in Northampton, Massachusetts, an institution only for women[2], as an associate professor. There she dedicated herself to teaching astronomy and making measurements and observations from the small Smith College Observatory. She often came to Boston to collaborate with the Harvard Observatory, which was much better instrumented, and where the most cutting-edge astronomy of the time was done. During her visits to Harvard she made friends with the new director of the observatory, Harlow Shapley, who had been appointed to this post a year earlier. In fact, her research life was mainly conducted at Harvard, given the lack of resources at Smith and the little support that research received there.

Her life remained largely unchanged for six years, teaching astronomy at Smith, making observations, mainly on stellar astrometry (measuring the position and movements of stars within the Galaxy, determining orbits in star systems, etc.), and publishing various articles in the *Bulletin of the Harvard Observatory*. In 1928, the third general assembly of the IAU took place in Leiden, Holland (which, as we saw, Mary Adela Blagg also attended). Priscilla had to save her own money to be able to attend the meeting, since her college, which gave more importance to teaching than to research, did not pay for her travel expenses. When she arrived, she got a surprise that would change her life.

---

[2] Smith's graduate programs now admit men; the college remains women only.

Nowadays, when a researcher attends a scientific meeting like this one, she has to go to the registration desk of the meeting (usually at the entrance of the venue where the sessions and conferences will take place), and present herself to receive the accreditation that identifies her as a participant in the meeting. But those were different times (and the scientific world was much smaller) and a welcoming committee was generally organized to pick up each participant individually at the train station to accompany her to the hotel, provide the accreditation badge, and take the participant to the place of the meeting. In general, doctoral students and younger researchers were part of these committees. The person appointed to receive Priscilla at the train station was a bright, young Dutch astronomer named Bart Bok (1906–1983), ten years younger than Priscilla, who was doing his doctoral thesis at Groningen. They met and got along very well and, despite the difference between their ages, Bok fell in love with Priscilla. They were together during the whole time of the general assembly, and before it was over, Bok, impetuous in character, asked Priscilla to marry him.

Harlow Shapley also attended this general meeting of the IAU. Priscilla introduced him to her young companion, and he must have impressed Shapley—so much that he tried to convince Bart Bok that he was wasting his time in Holland. He suggested that Bart change the subject of his doctorate research and go with him to the Harvard Observatory. There he would help him obtain a scholarship to finance his studies. That Bart was smitten with Priscilla, who also lived around Boston, was undoubtedly a weighty factor that inclined him to accept the proposal (the prestige that the Harvard Observatory had at that time would be another).

Bok accepted Shapley's offer. Priscilla, at first, did not do the same with Bok, since there was a large age difference between them, and she did not like the idea of living in Holland; besides, it was not clear if Bok could in fact come to Boston. But to her own surprise, she did bear feelings for him, so they maintained a correspondence for several months, waiting to see how events would unfold. That time was enough for Shapley to get Bok the promised scholarship, and for him to break with his thesis director. Finally, he arrived in the United States on September 7, 1929, to take this grand new step. Three days later, on September 10, Priscilla and Bart were married in Troy, New York.

Bart entered Harvard to work on his doctorate, while Priscilla was still doing her research at the Harvard Observatory without pay, while receiving her salary from Smith College. The following years were quite active. During the second year of Bok's doctorate, the couple's first child, John, was born. In 1931, she left Smith College for a job at Wellesley College, another university only for women, where she was better paid. The following year Bart defended his doctoral thesis "A study of the Eta Carinae region." In 1933, the couple's second child was born, a girl named Joyce Annetta, and Bart got a position as an assistant professor at Harvard.

Priscilla, on the other hand, had to settle for teaching jobs outside Harvard (and taking care of the children at the same time, which diminished her scientific productivity). As is evident in so many stories of the women of the moon, the observatories were patriarchal places where a woman had few opportunities to progress[3], and Shapley, as Pickering had done before him, economized to the maximum his personnel resources. He was delighted that a highly qualified astronomer like Priscilla worked for the observatory without his having to pay her a penny in return.

On the other hand, since she did not receive pay, nobody could demand that she follow a specific line of investigation. She worked as a free agent, and she researched and published what she wanted. But slowly Priscilla's lines of research converged with those of Bart, who was interested in the structure of the Milky Way. The Boks became great collaborators: he took the data and she analyzed it, she provided the necessary judgment and introspection to counteract his impetuous way of approaching new ideas, and together they discussed the meaning of the observations. As we read in an obituary for Bart Bok, "it is impossible, and sterile, to distinguish which were his achievements and which ones hers."

For almost thirty years, from 1929 until 1957, the couple remained based at Harvard[4] (she would eventually change her primary university again, going to teach at the Connecticut College for Women), co-authored numerous research papers, and worked on the shape of the Galaxy and the distribution of the stars in it. They also became involved

---

[3] Women in universities not only received lower pay than men, but there were also serious restrictions against hiring married women, as also happened with Cori Radnitz (whom we will meet in the next chapter). On the other hand, for men, being married was considered an advantage.

[4] They were very good friends to both Antonia Maury, whom they considered brilliant, and Annie Cannon, whom they considered an excellent person. And they found that there was no friendship between the two computers, not without reason: they were authors of rival stellar classification systems.

with the popularization of science, writing articles for the popular press and giving numerous public talks about their research (in 1936, *The Boston Globe* described them as "salesmen of the Milky Way"[5]). They were fascinated in particular by the region around the constellation of Carina, visible from the southern hemisphere but unobservable from Boston. Bart's thesis was about the curious system of the star Eta Carinae[6], whose mystery was unsolved at the time. In this same area it seemed that there was an arm of the Milky Way (the arm of Carina-Cygnus, today called arm of Sagittarius), and the Boks devoted much of their professional life to mapping this arm.

In 1937, the couple began the writing of what would become a famous textbook of astronomy entitled *The Milky Way*[7], which was first published in 1941, and eventually went through five editions. It became a tremendously popular book that was translated into numerous languages, and which also had a remarkable diffusion as a book of popular science, for its simple and enjoyable language. Each of the Boks wrote four chapters, divided according to each of their scientific interests, which they then traded to try to normalize their literary styles (which, incidentally, produced one of the few recorded disputes between the couple). But a complete uniformization of styles was not achieved. The astronomer Dirk Brouwer wrote a review about the book for the Sigma Xi society's magazine, in which he highlighted the chapters that he liked most and least; all the chapters that he liked most were those written by Priscilla.

The Boks also worked on the foundation, construction, and institutional startup of several astronomical observatories. In 1941, they spent several months in Mexico helping to start the National Mexican Observatory at Tonantzintla. In 1950, they went for a sabbatical leave to South Africa to install a new telescope at Harvard Observatory's new

---

[5] David Levy would borrow this phrase for the title of his book: *The Man Who Sold the Milky Way: A Biography of Bart Bok.*

[6] A star monster with a mass a hundred times that of the Sun, a blue giant that is thought to be ready to explode at any moment. It is such an unstable star that it spews large amounts of material into space in huge explosions, generating a striking nebula (that of the Homunculus, which dwarfs the Orion nebula) that surrounds the star; there was one particular explosion, detected in 1840, that is believed to have formed most of the current nebula. At that time, Eta Carinae was the second-brightest star in the sky, but since then the dust of the nebula has dimmed its brightness. In infrared, it is the brightest star in the sky.

[7] Bart Bok (1941). *The Milky Way.* Harvard University Press.

observation station in Bloemfontein[8], and in 1955, when the couple's interests had moved to radio astronomy, they studied a place to set up a radio observatory for the National Science Foundation. The conditions had to be such that it was relatively easy to get from Washington DC, in a valley that had minimal interference from planes, television, and other radio sources. The chosen place was Greenbank, West Virginia (where one of the authors did his doctoral research), nowadays one of the main observatories for radio astronomy of the world (and currently in danger of closure due to budget cuts)[9].

In addition to their efforts as planners and organizers, the Boks' scientific research continued. One of their main discoveries was to prove that star formation was not something that occurred only at the beginning of the history of the Galaxy (or the universe), but that stars are still forming today. The Boks saw that there were star clusters within our Galaxy that could not remain stable beyond ten million years; those stars, therefore, must have been formed at most ten million years ago— but the Galaxy was tens of *billions* of years old. Another discovery, and the one that has been most associated with them, are Bok globules: in 1946 they were struck by a curious type of small and compact dark nebulae of gas and dust, and they wondered if they were not seeing a star in formation. They proposed that those globular clouds, or "globules," as they called them, were hatcheries of stars, clouds in gravitational collapse within which increasingly higher temperatures were reached as the pressure in the center of the cloud increased, until the mass was hot enough to ignite nuclear fusion reactions and thus create stars. This point could not be verified until the nineties, when these globules were observed with infrared telescopes and it could be seen that, indeed, inside there were stars in formation.

In those days, a university was a place where tenured professors stayed working for life; leaving was not usual. And nothing seemed likely to indicate a change in the lives of the Boks; not even the Second World War was a serious break in their day-to-day lives. But world politics would

---

[8] This station was heir to Pickering's old station in Arequipa, Peru, funded by Catherine Bruce; in 1927 Harvard moved their southern-hemisphere operations to South Africa, due to the excess of cloudy days in Arequipa; Bloemfontein had much greater meteorological stability.

[9] Greenbank remains one of the few areas in the world where not only is there no mobile phone reception, mobile phones are in fact disallowed. It is blissfully quiet in both sonic and radio energies.

Figure 98 Dark clouds in IC 2944, a stellar nursery (https://www.eso.org/public/images/eso1322a/) ESO.

affect their working lives indirectly, after the war. The House Un-American Activities Committee, created in 1938, was beginning to accuse a large number of celebrities of being communists. One of the most active members of the committee in this "witch hunt" was Congressman John E. Rankin (1949–1953). He was a Democrat from Mississippi, a convinced racist (and possible member of the Ku Klux Klan), and a fierce anti-Communist who accused even Albert Einstein of being a communist. For Rankin, scientists were automatically suspicious, especially if they spoke publicly against the government's warmongering policy and the creation of nuclear weapons. One of those who came into Rankin's notice was Harlow Shapley, who was formally accused of communism and had to appear on November 14, 1946, before the Committee. Upon his return, after testifying before Congressman Rankin, Shapley said to the Boks in a broken voice: "That guy has the audacity to tell me I'm anti-American!"

The environment at the observatory became strained, and several colleagues turned against Shapley, which caused deep disgust to the

Boks. Both gave their public support to the director (in fact, Shapley had thought of Bart Bok as the next director after his retirement, and appointed him associate director), and Bart had to testify twice before the committee, which he remembered as one of the most unpleasant experiences of his life. One day they received a visit from a colleague at the observatory and close friend, Otto Struve (1897–1963), who brought a message from some of his colleagues: "I have been asked to come and tell you that your life would be much simpler at Harvard if you will stop supporting Shapley." As Bart said in an interview some time later, "that was the day we decided we should leave Harvard." This political pressure was compounded by scientific politics, as their colleagues at the observatory considered the couple's research on the Milky Way an unprofitable field, and thought that they should move to a more productive field of research.

Harlow Shapley was investigated by the FBI by orders of Commission for seven years, and for one year after he retired from the observatory in 1952. During that time all the correspondence he sent or received was intercepted and studied in detail. In a 1953 entry in the dossier of the FBI on Shapley (which unwittingly illuminates the intellect of these witch hunters), one can read: "Although the subject is no longer director of the Harvard Observatory, he continues to be active as a professor of astrology." Someone subsequently crossed out that word by hand and wrote "astronomy" instead.

After Shapley's retirement, the Boks lost all the support they could still muster at the observatory. Needless to say, Bart was not elected director. The new director was Donald Menzel (1901–1976), one of those who had turned against Shapley as a result of his accusation as a Communist. The couple received an offer from the University of Arizona, on the other side of the country. They liked the place and the researcher group, but they wanted a more radical change of air; they wanted to change country. More, they wanted to change hemisphere. As they told the group at the University of Arizona, their plan was to go for ten years to the southern hemisphere to thoroughly examine the Carina region and elucidate the structure of this spiral arm. Then, if Arizona continued to be interested, they would return. Their destination was Australia, where Bart had been offered the directorship of the Mount Stromlo Observatory. It was 1956.

Their stay in Australia was productive: they promoted a graduate program at the Stromlo Observatory, and the development of another observatory, that of Siding Spring, belonging to the Australian National University. Regrettably, they did not manage to completely unravel the structure of the Carina-Cygnus arm, since pursuing grants from the Australian government and the administrative tasks of the new observatory took a lot of time away from their research. But they did notable studies of the Magellanic Clouds (two small satellite galaxies of the Milky Way) and their interest in the region of the Eta Carinae nebula increased. It was a very happy time.

As they had promised, ten years later, in 1966, they returned to the United States and joined the University of Arizona, where Bart was appointed as director of the observatory (a post he held until his retirement). For Priscilla, who was already seventy years old, Arizona was instead a quiet retirement. Two years later, she would suffer a heart attack that would weaken her gradually in subsequent years. But that did not stop them from traveling and continuing to give lectures, and even briefly returning to Australia to visit their friends from Mount Stromlo in 1973.

In 1974, Bart, who despite being retired was vice president of the IAU, left this position to take care of the increasingly weak Priscilla at home. A year later, a meeting of Dartmouth, Michigan, and MIT took place in the Flandrau Planetarium of the University of Arizona, in Tucson, because these universities were going to build a joint telescope to place at the Kitt Peak Observatory. There were going to be three events associated with the meeting to which the Boks were invited: a dedicatory ceremony at the top of the observatory; a symposium in the planetarium on the identification of astronomical X-ray sources; and a big dinner the night before, to which a friend of the couple, Senator Barry Goldwater, was invited as a speaker.

The three invitations arrived separately and Bart asked his wife what they were going to do. Priscilla told him that she did not have the strength to drive to the Kitt Peak Observatory and stand there for who knows how long, and she also could not bear Senator Goldwater talking. On the other hand, he was ashamed to think that, with seventy years behind him and a life dedicated to astronomy, he knew nothing about astronomical sources of X-rays, so they decided they would go to the symposium in the planetarium.

On the day of the symposium, they arrived almost an hour before it began. To pass the time, Bart told his wife that they had installed in the planetarium a wonderful panorama of the Galaxy, the Milky Way, on which both had worked for so long. The display had small red lights that could be turned on and off to signal special characteristics of the Milky Way. Priscilla pressed one of the buttons and a faint light lit up to identify the Eta Carinae nebula. They were silent for a moment, remembering the wonderful photographs they had taken of this nebula during their years in Australia. Finally, Priscilla spoke: "You know Bart, when I leave, that's where I want to go. If I have to present myself soon to Saint Peter, I think I'll ask him to give me a front row seat in the center of the nebula. So I can see the stars forming right in front of my eyes. Look for me there, in Eta Carinae. I'll wait for you there." Four days later, Priscilla died of heart failure. At the Flandrau planetarium, in the chair in which she sat, a plaque with her name and the date of the symposium can be found.

Upon her death, Bart Bok established the Priscilla Fairfield Bok Prize to commemorate his wife's work. The award celebrates female American students who achieve the best academic results in the third year of their university studies in mathematics, physics, chemistry, computer science, geology, or applied statistics. There is a parallel Bart Bok prize for male students. The year after her death, the IAU honored her work by giving the name "Priscilla" to the asteroid 2137, and in 1979 baptized the Bok crater in honor of the couple.[10] (One wonders what happened to Joliot-Curie.) It is located on the far side of the Moon and is 45 km in diameter.

---

[10] There is another Bok crater on Mars, but it does not honor any person, referring rather to a people in Papua New Guinea.

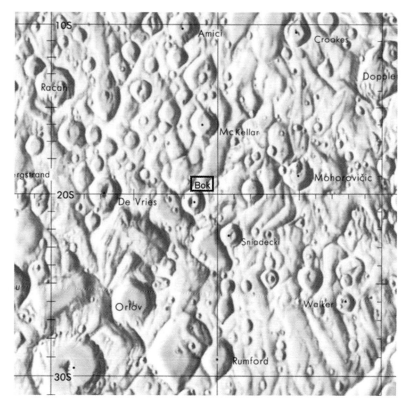

Figure 99 Location of crater Bok. Courtesy of the Lunar and Planetary Institute, Houston, Texas.

Figure 100 Lunar Reconnaissance Orbiter zoom on crater Bok (image width is 150 miles).

23

Gerty Theresa Radnitz Cori
(1896–1957)

Figure 101 Gerty Cori. Bernard Becker Medical Library, Washington University School of Medicine.

For all of us, Gerty was a human being of great spiritual depth. Modest, kind, generous and affectionate to a superlative degree and a lover of nature and art.

SEVERO OCHOA

Among the astronomers, physicists, and mathematicians on the Moon, Gerty Cori stands out. A Nobel Prize winner like some others, hers was in Medicine[1], in 1947. The Nobel committee awarded it to Carl Ferdinand Cori, Gerty Theresa Radnitz Cori, and Bernardo Alberto Houssay. The Coris took half for "their discovery of the process of the catalytic conversion of glycogen," while the other half was for Houssay (the first Latin American scientist—he was Argentine—to receive a Nobel Prize, and like the Coris honored with a Moon crater on the far side) for "his discovery of the role played by the hormone of the anterior lobe of the pituitary gland in the metabolism of sugar."

Gerty (from Gertrude) thus became the first woman to win the Nobel Prize in Medicine and the third woman in history to win any Nobel Prize for science after Marie Curie and Irène Joliot-Curie. She was born in Prague to a Jewish family, the eldest of three daughters of the chemist Otto Radnitz, who worked as a manager of a beet sugar refinery, and

[1] Formally, the Nobel Prize for Medicine or Physiology.

The Women of the Moon. Daniel R. Altschuler Stern and Fernando J. Ballesteros Roselló.
© Daniel R. Altschuler Stern and Fernando J. Ballesteros Roselló 2019. Published in 2019 by Oxford University Press. DOI: 10.1093/oso/9780198844419.001.0001

his wife Martha Neustadt. Up to the age of ten she was educated by private tutors, and following, in a school for girls. At sixteen, encouraged by her maternal uncle, who was a professor of pediatrics at the University of Prague, she decided to study medicine. She was able, over a couple of years of study, to learn Latin and the mathematics and sciences necessary to pass the requisite entrance exams (matura), and in 1914 she entered the medical school of the Karl-Ferdinands-Universität in Prague (today the Karlova University). Her story reminds us of the young Lise Meitner.

During her first year of study she found two loves: biochemistry and Carl Ferdinand Cori (1896–1984), a fellow student. In 1916, Carl put his studies on hold for a time when he joined the Austrian army's health corps on the Italian front during the First World War. Recall that Lise Meitner also went to the front to assist with the wounded, and for a time was on the Austrian–Italian front (while Marie Curie and her daughter Irène worked on the opposite front). We can imagine that the geodesics of Carl and Lise crossed at some point, but we do not know.

In 1920, after they had graduated from medical school, Gerty and Carl got married. In those difficult postwar years, she got a job in Vienna at the Caroline hospital for children, and he in a laboratory at the University of Gratz, after proving that he was not Jewish. The circumstances of the postwar period, including growing anti-Semitism, led them to seek employment abroad, and in 1922, Carl moved to the United States to take a job at the New York State Institute for the Study of Malignant Disease (now the Roswell Park Memorial Institute), in Buffalo, New York. Half a year later, when he was able to obtain a position as a pathologist for Gerty, she also emigrated. They formed a successful and solid research team. Although initially there was pressure against them collaborating (there were rules against couples working in the same department that usually affected married women), they ignored them and collaborated until Gerty's death.

The Coris investigated the regulation of glucose concentration in the blood, studying how sugar is metabolized (issues relevant to diabetes) and the role of lactic acid, hormones, and enzymes in the process. The metabolism of sugar is the source of energy for the activities of the body. Glycogen, a key stage in the mammalian energy system, is a polysaccharide (a chain of sugars) found in the liver and, at a lower concentration, in muscle tissues. In the liver, a hormone converts glycogen into glucose, which is transported by the blood to the muscles to be used as an energy source. Excess glucose returns to the liver transported by the blood in the form of lactate (lactic acid). The relationship between glyco-gen and glucose, regulated by hormones (insulin and epinephrine), is known as the Cori cycle, which they described in 1929[2].

During their years in Buffalo, Carl (but not Gerty) received several job offers from prestigious universities, which he rejected, since they did not accept work teams composed of a couple. Finally, in 1931 they accepted an offer from the Washington University School of Medicine in St. Louis, Missouri. He became the director of the pharmacology department and she started as a researcher, with a salary that was 10% of her husband's. There they established a laboratory that would become an important center for biochemical research. Their son Carl Thomas Cori was born in 1936. He studied chemistry and held various managerial positions at the Sigma-Aldrich company, a manufacturer of chemical products for research laboratories.

In 1941, as part of the war-induced wave of immigrants to the United States, the avant-garde laboratory of the Coris received Spanish bio-chemist Severo Ochoa (1905–1993), who after a year got a job at the

School of Medicine of New York University. In addition to Ochoa, others visited the Cori laboratory to study and collaborate with them. Several would receive Nobel Prizes later, including Ochoa in 1959: the English-Belgian Christian de Duve (1917–2013) in 1974, the Americans Arthur Kornberg (1918–2007) in 1959 (jointly with Severo Ochoa) and Earl Sutherland Jr. (1915–1974) in 1971, and the Argentine Luis Leloir (1906–1987) in chemistry in 1970.

The year 1947 brought the best and the worst news for the Coris. In October, they

Figure 102 Gerta and Carl in their laboratory in 1947. Smithsonian Institution.

[2] Carl F Cori and Gerty T. Cori (1929). Glycogen formation in the liver with d- and l-lactic acid. *Journal of Biological Chemistry,* Vol. 81, p. 402.

learned that they had won the Nobel Prize, news that they received with a remarkable lack of celebration, and to the astonishment of both their colleagues and interested journalists, they continued with their regular work routine. Shortly before their trip to Stockholm, the results of medical tests performed on Gerty, who had fainted a few months earlier, brought the bad news: she suffered from myelosclerosis, a deadly disease of the bone marrow. From then on, she would survive only with regular blood transfusions.

In his Nobel acceptance speech, Carl summarized the nature of the collaboration between them: "Our collaboration began 30 years ago, when we were still medical students at the University of Prague and has continued since then. Our efforts have been largely complementary, and one without the other would not have come as far as in combination."

Returning from Europe, they shared the prize money—$24,460.50—with some of their collaborators in the laboratory. The Nobel Prize improved Gerty's financial lot—as well as her status— in a more persistent way as well: she was appointed professor in the department of biochemistry at Washington University. In 1948, she also received the Garvan-Olin Medal, an award for American women chemists, given in recognition of her distinguished work in chemistry. She was named by President Harry S. Truman as a board member of the National Science Foundation, a position she held until her death, and she was elected to the National Academy of Sciences of the United States, the fourth woman to receive the honor.

In a 1950 radio program hosted by the well-known journalist Edward R. Murrow (1908–1965) entitled *This I Believe*, Cori discussed her values[3]:

My beliefs have undergone little change during my life, though I like to think they have developed into a somewhat higher plane. Honesty, which stands mostly for intellectual integrity, courage and kindness are still the virtues I admire, though with advancing years, the emphasis has been slightly shifted and kindness now seems more important to me than in my youth. The love for and dedication to my work seems to me to be the basis for happiness. As a research worker, the unforgotten moments of my life are those rare ones which come after years of plodding work, when the veil over nature's secret seems suddenly to lift, and when what was dark and chaotic appears in a clear and beautiful light and pattern.

---

[3] https://thisibelieve.org/essay/16457/ (including an audio).

During the ten years following her diagnosis, she carried the disease with courage and continued her work until shortly before her death in 1957 at sixty-one years of age. Her crater was approved by the IAU in 1979 and is on the far side. In this case there is no crater in honor of Carl (compensating somewhat for the exclusion of Irène Joliot-Curie). The United States Postal Service commemorated her work as well, as one of a series of four stamps in honor of distinguished scientists issued in 2008 (Gerty Cori, Linus Pauling, Edwin Hubble, and John Bardeen). Few noticed that the formula on Cori's stamp contains an error in the location of the link between the glucose molecule and the phosphate group OPO_3. But after all, it is a commemorative stamp and not a biochemistry text.

In 1957, Bernardo Houssay spoke at a ceremony in Gerty's memory[4]:

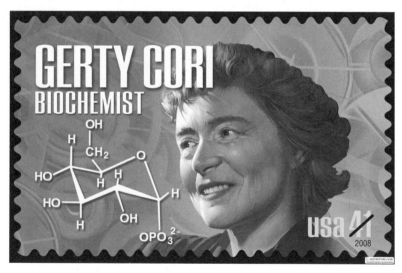

Figure 103 Commemorative postage stamp of Gerty Cori's work (left) and the correct α-D-glucose 1-phosphate molecular (right). USPS.

[4] Bernardo A. Houssay (1957). *Memorial to Gerty Theresa Cori*. The Bernard Becker Medical Library.

Gerty Cori's life was a noble example of dedication to an ideal, to the advancement of science for the benefit of humanity. The work of the Coris has permanent value and has already led to fundamental discoveries in the knowledge of cell physiology. Gerty Cori's charming personality, so rich in human qualities, won the friendship and admiration of all who had the privilege of knowing her. Her name is engraved forever in the annals of Science and her memory will be cherished by all her many friends as long as we live.

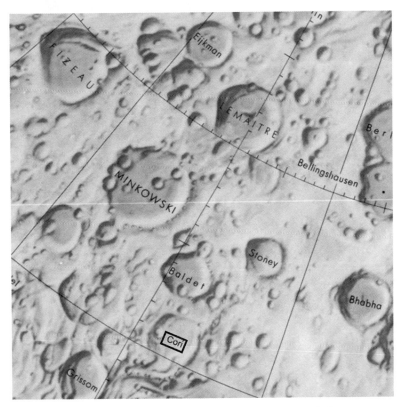

Figure 104 Location of crater Cori. Courtesy of the Lunar and Planetary Institute, Houston, Texas.

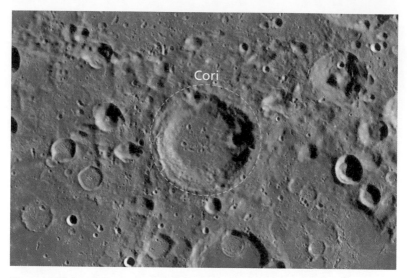

Figure 105 Lunar Reconnaissance Orbiter zoom on crater Cori (image width is 150 miles).

24

Judith Arlene Resnik (1949–1986)

Figure 106 Judith Resnik.
Courtesy of NASA.

I think astronauts probably have the best jobs in the world.

JUDITH RESNICK
at a presentation in 1985

Judith Resnik was one of the two female astronauts who died in the Space Shuttle Challenger catastrophe in 1986. Born in 1949 in the town of Akron, Ohio, she was the daughter of Jewish immigrants. Her father, Marvin, was an optometrist, and her mother, Sarah, a court clerk, both originally from Ukraine. She also had a brother, four years younger, named Charles.

She quickly showed herself to be a child prodigy, already reading and solving simple mathematical problems at the age of three, so she skipped kindergarten and entered directly into primary education. Her father, with whom she was very close all her life, liked engineering and taught her to make simple machines and electrical work, instilling in her a vocation for engineering. Her relationship with her mother was not so intimate; Sarah was a rigid woman who believed in organization, in discipline, and in strict compliance with schedules. Nonetheless, she taught her to play the piano, an art in which Judith excelled during her teenage years and which she even considered as a profession. But she was more attracted to mathematics and science, subjects in which she was particularly gifted.

Her classmates and teachers described her as an extremely bright student. Her teachers reported that she was "more serious than her

The Women of the Moon. Daniel R. Altschuler Stern and Fernando J. Ballesteros Roselló.
© Daniel R. Altschuler Stern and Fernando J. Ballesteros Roselló 2019. Published in 2019
by Oxford University Press. DOI: 10.1093/oso/9780198844419.001.0001

classmates, but very warm and friendly." She had excellent grades, as a disciplined perfectionist intensely focused on her work. At her secondary school in Akron, she joined the chemistry, French, and math clubs (she was the only female member of the last), and she participated in numerous after-school activities, probably as an escape from the increasingly frequent arguments between her parents, who ended up divorcing in 1966, the same year Judith finished high school, first in her class. Initially, she tried to live with her mother, but after several disagreements, she went to the judge to transfer custody to her father (who would remarry shortly afterwards).

After high school, and after definitively discarding the piano, she enrolled at Carnegie Tech University (now Carnegie Mellon), where she majored in electrical engineering. There she met another engineering student, Michael Oldak, whom she married as soon as they graduated, in 1970. The couple moved to New Jersey, where they were both employed by the Radio Corporation of America (RCA). Judith worked in the missile and radar division, in which components of radio transmission and reception were developed, mainly subcontracted by the military, but also for space projects. She combined this work with night classes at the University of Pennsylvania in Philadelphia, earning a master's degree in engineering.

After a year working in industry, Michael was accepted at Georgetown University in Washington DC to study law, which meant a radical change to his academic life. So, the couple moved to Washington DC, and Judith was transferred to the local division of the RCA. She transferred to the nearby University of Maryland, where she completed her master's degree in engineering. Later, she decided to continue her studies towards a doctorate in engineering at Maryland. Something in this change of environment did not sit well with the couple, because in 1975, after five years of marriage, they divorced, for reasons not entirely clear (Judith was always reserved about the subject, even with her parents, although a close friend of hers, Connie Knapp[1], blames the separation of the young couple to the fact that he wanted to have children and she, with her doctorate half done, did not). Despite this, it was an amiable separation and they continued to be friends.

In 1977, when Resnik was completing her doctorate, NASA began to focus its recruitment program on women and minorities, and she

---

[1] Elizabeth Kolbert (1986). *The New York Times*, February 9.

decided to apply. Keeping her options open, as soon as she got her PhD, she started working at Xerox as a systems engineer, but to her surprise she was accepted into NASA's astronaut training program in 1978, along with five other women (out of a total of thirty-five admitted). Needless to say, she immediately agreed and left her job at Xerox to move to NASA's Johnson Space Center in Houston, Texas. There she enjoyed the atmosphere of serious and rigorous work that so well matched her character, and received the nickname of "J. R." by which she would always be known in the astronaut community.

After six years of training, she finally was offered the post of specialist to handle the robotic arm aboard the space shuttle Discovery on its inaugural mission in 1984. She was the only woman of the six astronauts on board, a group of fresh astronauts (for all but one of them, this was their first trip to space) who received the name "zoo crew." Judith returned delighted and commented on her decision to become a professional astronaut "for as long as NASA wants me." This mission made her the second American woman to go into space; the first had been Sally Kristen Ride (1951–2012), a classmate from the same recruiting batch as Judith, who had flown into space the previous year on board Challenger.

Challenger would also be the next shuttle ride for Resnik, only two years after her first flight. Its mission in 1986 had two main objectives: to put into orbit a communications satellite for NASA and to take spectral data from the tail of Halley's Comet. Judith's job as specialist would be to take the spectra of the comet. As is well known, none of this came to pass. The space shuttle Challenger exploded spectacularly two minutes after launch, killing all seven crew members in one of the major catastrophes in the history of astronautics.

This disaster led to the immediate shutdown of the shuttle program for two years while the causes of the accident were investigated and the necessary measures were established in order to avoid further tragedies[2]. The extremely complex shuttle system, a machine composed of millions of pieces, had failed because of a simple rubber O-ring that was too cold. The circumstances leading up to the tragedy show a disastrous collision between physics and human institutions. The launch of the mission had been postponed five times due to bad weather conditions.

---

[2] These measures were not completely effective, as years later the Columbia disaster would show.

On this occasion there were also recommendations from the engineers to postpone the launch, due to the previous night's low temperature, which had caused ice to form at the launch tower. But perhaps due to the fact that the mission was under heavy media attention (the first teacher-astronaut was on board, as we will see in the next chapter), it was thought that five postponements were enough. The rubber O-rings that sealed the joints between the different sections of the solid-fuel side rockets, so characteristic of the space shuttles, were intended to seal these sections and prevent leakage of the high-pressure gases produced by combustion. But the correct functioning of these rubber O-rings was guaranteed for ambient temperatures above 12°C and Morton Thiokol, the manufacturer, did not ensure a proper seal for lower temperatures. However, on the cold night before the launch, the joints were chilled to 4°C.

The Nobel Prize-winning physicist Richard Feynman (1918–1988), a member of the commission that investigated the catastrophe, demonstrated on television how the rubber of the joints lost its elasticity when it cooled, simply submerging a ring of this material in ice water[3].

Takeoff and atmospheric reentry are the two most dangerous moments of a space flight. Takeoff is basically sitting on top of the (relatively) slow and (relatively) controlled explosion of a powerful bomb, with all the risks that this entails. To achieve a stable orbit, very high speeds must be reached, leaving very little room for error.

Since takeoff begins with zero speed, the spacecraft must accelerate at each point of its ascent. While speed is increasing with altitude, the air density that the craft is traveling through is decreasing: the atmosphere at sea level is denser and it becomes more tenuous as the ship ascends, until it is almost nonexistent at the altitude at which the shuttles orbit. Between launch and orbit there is a moment when the atmosphere is still dense enough and the speed of the ship is already so high that, if nothing attenuates it, the ship could suffer serious structural damage due to the high dynamic pressures. In astronautics, this zone is called Max Q, the zone of maximum dynamic pressure. This pressure is usually mitigated by reducing the power of the engines slightly in order to lower the dynamic pressure. Through this process, the craft continues

---

[3] Feynman was also very critical of NASA, and its highly unrealistic risk management policy (NASA estimated the probability of failure as 1 in 400). He concluded that, "for a technology to be successful, reality must have priority to public relations, because nature cannot be fooled."

to rise by inertia, until the Max Q region is crossed. At the moment when Max Q has been passed (for the Space Shuttle, at about 50 seconds into the flight), the instruction is automatically given to return the engines to maximum power and inform the crew of this with: "Go at throttle up." It is a moment of relief for the astronauts that indicates that the most dangerous part of the ascent has passed. In a sad irony, the last transmission received from the shuttle was the voice of its commander, Dick Scobee, confirming the receipt of this command: "Roger, go at throttle up."

On board the Challenger that January morning, the failure of the rubber gasket led to high-pressure flames leaking out of the side of the booster rocket. This flame was aimed like a torch at the Shuttle's huge external fuel tank, and it burned through the wall of the tank, releasing all its liquid hydrogen and liquid oxygen propellants. As the chemicals mixed, they ignited to create a giant fireball. The shuttle itself, however, was still intact at this point and still rising. It was at an altitude of 48,000 feet.

Apparently, some astronauts survived the explosion. The cockpit, of greater strength than the rest of the ship, was torn off from the rest of the disintegrating shuttle and remained intact. It arced upward, then fell, taking three minutes from the detonation before hitting the ocean at over 200 miles per hour. At the moment the Shuttle was blown apart, the cabin suffered an acceleration of about 15 g (that is, 15 times that felt on the earth's surface), which after two seconds had already decreased to 4 g, and after ten seconds to 0 g, free fall; accelerations that could have been borne by the astronauts. The astronauts' personal air supplies, which provide up to six minutes of oxygen for emergencies, were recovered after the catastrophe and three were found to have been manually activated after the explosion. One of them was Judith Resnik's own, another, that of specialist Ellison Onizuka, sitting next to her, and another, that of the pilot, Michael Smith, who was sitting in front of Resnik and whose oxygen supply had to be activated from behind— either Resnik or Onizuka was the one who activated it. In addition, several electric switches in front of the pilot were changed from their launch positions, probably in a futile attempt to recover the electrical power supply. At least three of the seven astronauts were still alive after the ship's disintegration, but there was no way out of the trap that the cabin had become, there were no ejection seats with a parachute or even the possibility of opening the cabin with pyrotechnics. NASA considered that the reliability of the space shuttle was so high that no escape system was necessary; the cost of such a system was not justified.

During Judith's memorial, then-senator and former astronaut John Glenn (1943–2016) said that the last words received from the Challenger, "Go at throttle up," perfectly defined Resnik's personality and life. Her father remarked in turn that "she had the brain of a scientist and the soul of a poet." In the year of her death, the Institute of Electrical and Electronic Engineering instituted the Judith A. Resnik Award for those persons or teams with exceptional contributions to the field of space engineering. Her name was given to asteroid 3356. And two years after the catastrophe, in 1988, the IAU baptized seven craters on the Moon with the names of each of the astronauts killed in the Challenger catastrophe. All of them are located inside the huge Apollo crater, on the far side of the Moon, very close to one another. The Resnik crater is a small crater of rounded, worn shape, holding in its interior, to the northwest, a smaller crater, which gives the appearance of an eye looking up to the left.

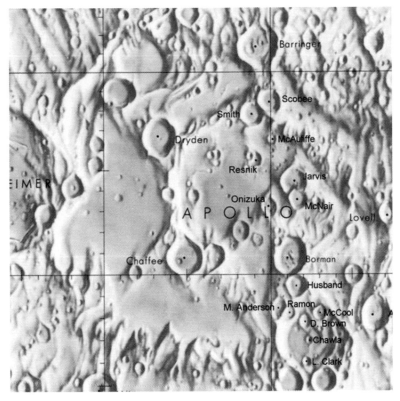

Figure 107 Location of craters named after Space Shuttle astronauts.
Courtesy of the Lunar and Planetary Institute, Houston, Texas.

Figure 108 Crater group, in context with the Apollo basin and craters there named after Space Shuttle astronauts. The larger crater at center left, named in honor of Apollo 1 crew member Roger Chafee, is roughly 31 miles across. NASA/USGS/ASU.

25

Sharon Christa McAuliffe (1948–1986)

Figure 109 Christa McAuliffe.
Courtesy of NASA.

I haven't felt frightened, now I'm not sure how I'm going to feel when I'm sitting there waiting for takeoff and those solid rocket boosters ignite underneath me and everything starts to shake. But right now, I think instead of being apprehensive I'm just very excited about doing it.

CHRISTA MCAULIFFE (1985)[1]

Christa McAuliffe was the other woman who perished in the Challenger disaster. A high school teacher, she would have been the first civilian to go into space after President Reagan announced[2]: "I'm directing NASA to begin a search in all our elementary and secondary schools and to choose as the first citizen passenger in our space program, one of America's finest, a teacher . . ." Born in Boston, Massachusetts, in 1948, she grew up in the suburb of Framingham, and was the oldest of five siblings. Her father, Edward C. Corrigan (1922–1990), was an accountant, and Grace, her mother, was a substitute teacher, likely the inspiration for Christa's vocation. As a child she was more mature than expected for her age. She was energetic and determined, a good student with clear ideas, and dedicated to her objectives. While in school, she also studied piano, and even gave concerts[3]. She was a Girl Scout and in high school she stood

---

[1] Quote taken from an interview in *USA Today*.
[2] Grace George Corrigan (2000). *A journal for Christa: Christa McAuliffe, teacher in space*, p. 97. University of Nebraska Press.
[3] As we will see, there are some other parallelisms with the life of Judith Resnik.

The Women of the Moon. Daniel R. Altschuler Stern and Fernando J. Ballesteros Roselló.
© Daniel R. Altschuler Stern and Fernando J. Ballesteros Roselló 2019. Published in 2019 by Oxford University Press. DOI: 10.1093/oso/9780198844419.001.0001

out as a softball star. She was still in high school when she met her future husband, Steven McAuliffe.

After high school, Christa enrolled at Framingham State College, where she pursued a degree in history and education, while her boyfriend Steven studied at the Virginia Military Institute. Like Judith Resnik (and the same year, 1970), as soon as Christa graduated, she and Steven got married. After the wedding they went to live in Maryland, near Washington, DC, so that Steven could attend the Georgetown University Law Center (he was, therefore, a classmate of Judith's husband). Christa began her teaching career in Maryland, instructing high schoolers in American history and English for seven years, while pursuing post-graduate education at Bowie State University. She earned her master's degree in education supervision and administration in 1978. A year later, the couple moved to Concord, New Hampshire, where he had taken a position as assistant district attorney, and she continued working as a teacher. In the meantime, they had two children, Scott and Caroline.

Nothing seemed to indicate that her life would be very different from that of any other high school teacher. But in 1984, President Ronald Reagan announced the "Teacher in Space" program, which would change her life and, eventually, take it. This program aimed to put a teacher in space, a civilian aboard the Space Shuttle, to "inspire students, honor teachers, and stimulate interest in mathematics, science and space exploration." The program looked for a brilliant educator, someone who could communicate with enthusiasm and efficiency, since this teacher would have to teach from space. As soon as Christa heard about this initiative, a new horizon opened up, as she was an enthusiast of space adventure, although she had never dreamed of being part of it. Without hesitation, she sent her application to the program, one of 11,500 applicants. In her application she wrote[4]:

> I remember the excitement in my home when the first satellites were launched. My parents were amazed and I was caught up in their wonder. In school, my classes would gather around the TV and try to follow the rocket as it seemed to jump all over the screen. I remember when Allan Shepard made his first historic flight – not even an orbit – and I was thrilled. John Kennedy inspired me with his words about placing a man on the moon and I still remember a cloudy, rainy night

[4] Framingham State University (1985). *Teacher in Space Application - Christa McAuliffe.* Challenger STS-51L: Ephemera. Book 19. http://digitalcommons.framingham.edu/challenger_ephemera/19

driving through Pennsylvania and hearing the news that the astronauts had landed safely.

As a woman, I have felt envious of those men who could participate in the space program and who were encouraged to excel in the areas of math and science. I felt that women had indeed been left outside of one of the most exciting careers available. When Sally Ride and other women began to train as astronauts, I could look among my students and see ahead of them an ever-increasing list of opportunities.

I cannot join the space program and restart my life as an astronaut, but this opportunity to connect my abilities as an educator with my interests in history and space is a unique opportunity to fulfill my early fantasies, I watched the Space Age being born and I would like to participate.

Because of her letter of application, she made it through the first selection round, in which one hundred and fourteen teachers (one percent of the candidates) advanced to a second round in Washington, DC. Here, they pressed their candidacy to a jury composed of former astronauts, scientists, teachers, politicians, and others. This jury selected ten of the one hundred and fourteen. These ten would go on to perform the tests to see that they had the necessary physical and psychological capacities needed. The final selection was made by seven high-ranking officials from NASA. Christa addressed this group: "I have always had the feeling that ordinary people have not been given their place in history. I want to humanize the space age by giving it the perspective of a non-astronaut. Space is the future. As teachers, we prepare students for the future. We have to include it, space is for everyone." Unanimously, the court chose Christa as "first teacher in space."

In her memoirs Christa's mother writes[5] that on the morning before the announcement of the committee's selection, her husband Ed remembered one of Christa's favorite songs: André Previn's "You're Gonna Hear from Me." He thought it was a good omen:

> Move over sun and you, give me some sky
> I've got some wings, I'm eager to try
> I may be unknown but wait 'til I've flown
> You're gonna hear from me

On July 19, 1985, Vice President George Bush, in a ceremony with the ten finalists, announced the election of Christa McAuliffe as the first

[5] Grace George Corrigan (1993). *A journal for Christa: Christa McAuliffe, teacher in space.* University of Nebraska

civilian passenger in the history of space flights: "The President said last August that this passenger would be one of America's finest—a teacher. Well, since then, as we've heard, NASA, with the help of the heads of our State school systems, has searched the Nation for a teacher with 'the right stuff.' There are really thousands [. . .] Today we honor all those teachers here."

At the end of the summer of that year, Christa began three months of training as an astronaut at NASA's Johnson Center, where she would meet her flight partner, Judith Resnik. There she learned to maneuver in a spaceship, sleep, eat, wash, and go to the toilet in space, to take pictures clearly (as she was going partly as a space flight documentarian), and even to fight fires and play a role in other emergency procedures. As a teacher in orbit, Christa also prepared two classes about space, to be taught from the Space Shuttle by live broadcast to American schools. During the months prior to the mission, Christa participated in numerous national television programs, including the Tonight Show, then hosted by Johnny Carson. Asked if she felt nervous about her appearance she laughed and replied[6]: "I've handled children for fifteen years in the classroom; I can manage fifteen minutes with Johnny Carson." All along, she championed the work of teachers, building public support for the profession as a whole, and aiming to increase the vocation for teaching among students. The "Teacher in Space" program became wildly popular.

Because of this, media coverage of the Challenger flight was much higher than usual. When disaster struck, the attention of the entire country—and of the whole world—was already turned on the launch. Millions who watched (either live or one of many recordings broadcast all over the world) were horrified by what they witnessed, unable to do anything. We authors, from different parts of the world, saw the takeoff and the subsequent explosion which stunned us. Inconceivably greater was the pain and grief of the families of the crew members, who watched in terror as their loved ones were leaving this world.

Christa left a husband and two small children (Scott was eight years old and Caroline was five years old—both have become teachers). Steven, her husband, depressed and furious, lived secluded with her children in Florida for a year. Even today, over thirty years later, her family and

---

[6] Grace George Corrigan (1993). *A journal for Christa: Christa McAuliffe, teacher in space*, p. 109. University of Nebraska.

friends refuse to make statements to the press. As more details were discovered about the accident, its avoidable cause, and the crew's excruciating final minutes, anger replaced grief in the McAuliffe family. Christa's father wrote in his diary[7] (published after his death in 1990 by his wife):

> My daughter Christa McAuliffe was not an astronaut – she did not die for NASA and the space program – she died because of NASA and its egos, marginal decisions, ignorance, and irresponsibility. NASA betrayed seven fine people who deserved to live. One of the Commissioners stated, "It was no accident, it was a mistake. [...] They [NASA] deliberately neglected to make the necessary corrections to the O-rings and are, therefore, as guilty as if they planned to deliberate criminal act."

Christa's mother, on the other hand, although agreeing in part, holds no grudge against NASA: "It's true, they underestimated the risks, but people are human, they did not deliberately send these people off to their death."

After the disaster, Barbara Morgan, who came second in the "Teacher in Space" project, assumed Christa's lead role in the program[8], but it was finally cancelled due to the accident. Christa had boldly stated before the NASA selection committee that ordinary people had not been given a place in history. But she did (tragically) go down in history. Today some forty schools and institutes around the world are named after her. In addition to the seven lunar craters mentioned in the last chapter, the IAU decided to baptize asteroids 3350 to 3356 with the names of the deceased crew members; the asteroid 3352, discovered in 1981, is the one assigned to McAuliffe. Her crater, inside Apollo, on the far side of the Moon, is next to that of Resnik, to the northeast. It is the smallest of the seven dedicated to Challenger's crew.

---

[7] Grace George Corrigan (1993). *A journal for Christa: Christa McAuliffe, teacher in space.* University of Nebraska.

[8] Morgan would go on to become a professional astronaut, and fly with the shuttle Endeavor to the International Space Station in 2007.

26

Kalpana Chawla (1962–2003)

The path from dreams to success does exist. May you have the vision to find it, the courage to get on to it, and the perseverance to follow it.

KALPANA CHAWLA[1]

This item is not even worth mentioning other than wanting to make sure that you are not surprised by it in a question from a reporter. During ascent at approximately 80 seconds, photo analysis shows that some debris from the area

Figure 110 Kalpana Chawla. Courtesy of NASA.

of the -Y ET Bipod Attach Point came loose and subsequently impacted the orbiter left wing, in the area of transition from Chine to Main Wing, creating a shower of smaller particles. The impact appears to be totally on the lower surface and no particles are seen to traverse over the upper surface of the wing. Experts have reviewed the high-speed photography and there is no concern for RCC [reinforced carbon steel] or tile damage. We have seen this same phenomenon on several other flights and there is absolutely no concern for entry. That is all for now. It's a pleasure working with you every day.

> Message of January 23, 2003, from flight director STEVE STICH to commander Rick Husband of Columbia

Mission STS 107 of the Columbia Space Shuttle (the oldest shuttle) took off from the Kennedy Space Center in Florida on January 16, 2003. It was the twenty-eighth time that Columbia flew; its inaugural flight (STS 1) had been on April 12, 1981. Later, with Columbia already in orbit, the

---

[1] Excerpt from an email she had sent to students at her former college, the Punjab Engineering College. https://twitter.com/NASAIVV/status/1003599392279027712

The Women of the Moon. Daniel R. Altschuler Stern and Fernando J. Ballesteros Roselló.
© Daniel R. Altschuler Stern and Fernando J. Ballesteros Roselló 2019. Published in 2019 by Oxford University Press. DOI: 10.1093/oso/9780198844419.001.0001

routine analysis of monitoring camera footage showed that 82 seconds after takeoff, at a height of 12 miles and a speed of almost one mile per second, a piece of insulating foam the size of a briefcase, used to prevent the formation of ice on the outside of the main fuel tank came off and hit the lower part of Columbia's left wing. After the impact, a rain of whitish particles was visible, flying off the wing. The nature and location of the possible damage was not clear, and engineers asked their NASA managers to investigate thoroughly any potential damage. But they were not alarmed, they thought that the impact had not caused critical harm, which could come to a head during the dangerous process of reentry into the atmosphere. In the past there had been similar incidents of impact to the fragile thermal protection tiles of the shuttle without catastrophic failure, and it was understood that the absence of a tile or two could not cause the loss of the vehicle. These assumptions were wrong.

Like takeoff, reentry is one of the most dangerous moments of these missions (for an airplane the case is similar, though less dramatic). During reentry, the shuttle must get rid of the large amount of kinetic energy associated with orbital movement: an object of seventy-five tons in orbit at 200 miles above the earth and descending at a speed of 17,000 mph, must land at only about 180 mph. As it enters the atmosphere, the ship is slowed by friction and its kinetic energy (proportional to its mass and the square of its speed) is transformed into heat. The shuttle's structure must withstand temperatures of up to 3000°F (1650°C). To prevent the shuttle burning up, it is covered with special tiles of different compositions (thermal protection system) that act as thermal insulation.

The loss of a significant number of tiles can be lethal. The reentry maneuver is so delicate that the ship must enter the atmosphere at a very specific angle with a tolerance of only two degrees. If the ship enters at a greater angle than seven degrees, it passes through layers of very dense atmosphere too quickly, causing such excessive heating that even the tiles can't protect it. On the contrary, if it enters with a smaller angle than five degrees, the ship can bounce against the atmosphere, like skipping a stone across a pond.

Returning from its successful sixteen-day mission on February 1, 2003, Columbia crossed the California coast at 08:53 at an altitude of about 40 miles. Traveling at about 17,000 mph over the ground, it was 23 minutes away from its expected landing in Florida. A recovered video[2]

---

[2] Available on YouTube.

allows us to see the activity in the cabin as the crew prepares for reentry: we listen to their voices and see the astronauts putting on their gloves, greeting the camera, and checking the systems for reentry and landing. Upon reentering, they comment on the yellow and orange light effects observed through the windows as the heat from the shuttle turns the air outside into plasma through the heat and pressure of its passage. You hear the comment: "it looks like a blast furnace", and laughing: "you would not want to be out there now".

On the ground, photographers observe a large red flare that came off the ship instead of the expected plasma trail. "Wow," exclaims one, "Have you seen that? Something came off the shuttle!" It was a vision they would not forget. Many of those watching and photographing the shuttle witnessed distinct events as debris came off the orbiter. Seconds later, at 08:55, mission control noticed that the temperature and pressure sensors began to indicate anomalous data from the left wing. The telemetry of several temperature sensors stopped working. Seconds later the pressure telemetry of the landing gear tires of the left wing was lost.

Mission control sent a message: "Columbia, Houston, we see your tire pressure messages and we did not copy your last" (referring to the anomalous measurements in tire pressure on the left landing gear). The answer from Rick Husband, Columbia's commander, at 08:59:28: "Roger uh bu . . ." (Roger is used in messages to indicate "understood", or "received"). Then, silence.

Suddenly the routine atmosphere—tense but capable—in mission control changed to a high level of anxiety. Something was clearly, horribly wrong. "Columbia, Houston, comm check" aired several times without an answer. The radar units that expected to detect the Columbia did not. Meanwhile, alarms sounded in Columbia's cabin. Columbia was disintegrating over the state of Texas. Sixteen minutes from landing in Florida, the cabin began to shake violently while rapidly losing pressure, and the astronauts fell mercifully unconscious.

On this occasion, the loss of some protection tiles proved fatal. It allowed the left wing to overheat to such an extent that it lost its structural integrity. An image obtained from New Mexico showed material emerging from the left wing of Columbia. Seventeen years after the Challenger disaster, another seven astronauts perished on a shuttle, again including two women: Kalpana Chawla and Laurel Blair Salton Clark.

Kalpana Chawla[3]—her family called her "Montu"—was born in the small town of Karnal, of about 300,000 inhabitants, on March 17, 1962, in the Indian state of Haryana, one of the richest states of India, in the far north of the country. Kalpana was the youngest of four children, the third daughter. At that time, the birth of a son was a cause for celebration, that of a daughter reason for a silent disappointment, something that is still common in many places in this world.

Her life, until her tragic death at age 41, is the story of a girl who was often told "no" simply because she was a female, but who refused to accept things that way, a quiet but determined rebel. Although her father, a successful industrialist, was conservative, her mother and her brothers supported Montu's defiance.

When she turned three, she chose the name Kalpana, which means "imagination," as her formal name. She graduated from the Tagore School in Kamal in 1976. From an early age Chawla knew she wanted to be an aerospace engineer, inspired by the aircraft of a local flying club, and she even got her father to get her to fly over the plains of Haryana. Flying was her passion. Her younger brother accompanied her the day she applied to be admitted to the Punjab Engineering College in Chandigarh to study aeronautical engineering. She was one of the first four women graduates in engineering there, obtaining her degree in 1982.

Her desire to progress was insatiable; she applied and was accepted to the graduate aeronautical engineering program at the University of Texas at Arlington, which she entered in September 1982, against the wishes of her family, especially her father. There she met Jean-Pierre Harrison, a pilot with whom she began flying planes and whom she married in December of 1983. She earned a master's degree in 1984 and then the couple moved to Boulder, Colorado, where Kalpana continued her stud ies at the University of Colorado, receiving her doctorate in 1988.

After finishing her studies, she started working for NASA's Ames Research Center in California, doing computational fluid dynamics. She became a United States citizen in 1991, and in 1993 she was hired as vice president and scientific researcher at a company in Los Altos called Overset Methods. But in December 1994, she took a huge turn towards her goal when she was called to an interview for the astronautics pro-

[3] Jean-Pierre Harrison [her husband] (2011). *The edge of time: the authoritative biography of Kalpana Chawla*. Harrison Publishing.

gram at NASA's Johnson Space Flight center in Houston. After rigorous exams, she was accepted to the program in December of 1994.

She learned to pilot military aircraft and to parachute, complementing her civilian and engineering expertise, and was finally assigned to the crew of the Columbia for mission STS 87, taking off on November 19, 1997, and returning on December 5. The launch was observed at the Kennedy Space Center by several relatives including her father, who had traveled from India. Kalpana flew high. Thirteen years later, she was again selected to go to space, for the mission STS 107, news that filled her with joy, and on the day of the launch her family again traveled to witness it. She did not return. Her death certificate indicates that the place of death was the "airspace over Texas."

Asteroid 51826 Kalpanachawla, is one of seven that were named in memory of the Columbia astronauts. Her crater located on the hidden face of the Moon is next to crater L. Clark, her astronaut companion, whom we will meet next.

27

Laurel Blair Salton Clark (1961–2003)

Figure 111 Laurel Clark.
Courtesy of NASA.

We're looking at Earth science, observing our planet. Also, space science, looking at the ozone in the atmosphere around our Earth. Also looking at life science. And on a human level, using ourselves as test subjects.

LAUREL CLARK (2003)

Laurel Clark was the other woman casualty of the Columbia catastrophe. Born in Iowa in 1961, she had an itinerant childhood, passing through New York, New Mexico, and Missouri (among others), until finally her family settled in Racine, Wisconsin, where she lived most of her life. Her companions and family members remember her as a well-rounded person with an adventurous spirit and multiple interests, such as skiing, swimming (she often worked as a lifeguard), climbing (once climbing Mount Fuji in Japan), and parachuting.

During her high-school studies, she was attracted to biology, but one of her professors turned her interest more specifically toward medicine. She began her medical studies at the University of Wisconsin-Madison, and in order to be able to pay for them, she enlisted in the Navy. Laurel once remarked that joining the military was something she did purely for financial reasons, but that once she was inside, the navy "hooked" her because of the variety of opportunities it offered.

In 1987, she obtained her doctorate in medicine and began her training as a navy medical officer. She specialized in submarine rescue and medicine under immersion conditions, and made several evacuations of submarines; undoubtedly, a qualitative leap with respect to her

The Women of the Moon. Daniel R. Altschuler Stern and Fernando J. Ballesteros Roselló.
© Daniel R. Altschuler Stern and Fernando J. Ballesteros Roselló 2019. Published in 2019
by Oxford University Press. DOI: 10.1093/oso/9780198844419.001.0001

youthful experiences as a lifeguard in swimming pools. After that, she trained as an aviator, and finally obtained the title of naval flight surgeon and the rank of commander. During diving classes, she met the love of her life, another naval doctor named Jonathan Clark, whom she would marry in 1991.

In 1994, NASA became interested in her and she was interviewed as a possible candidate for astronaut. It was unfortunate that she was at that time pregnant with her son, Iain, which meant she was excluded from the program. However, she remained interested in space exploration and she tried her luck again the following year, among another 2400 applicants. Of them, only forty-four were selected, and she was one of them. In this 1996 astronaut promotion, almost twenty years after Resnik and Ride, about a quarter of the selected candidates were women.

She began the long and hard training to become an astronaut. During this period, her husband Jonathan also joined NASA, in 2001, as a flight surgeon and specialist in neurology and space medicine[1]. As part of her training at NASA, Laurel went to Russia to learn how to use the Russian space suit Sokol and the Soyuz systems, as she was initially selected for a long-term mission on the International Space Station.

However, her only mission in space was as a specialist aboard STS 107. A mission dedicated to pure research and to carrying out scientific experiments (around eighty), mainly related to biology and growth in conditions of weightlessness. Laurel would take several samples of her blood, saliva, and other fluids to see how space affected the human body, and act as a "space farmer", growing roses and other plants. Seeing how a silkworm made its cocoon in weightlessness and then the butterfly emerged, she excitedly commented by radio to a CNN journalist[2]: "Life continues in lots of places, and life is a magical thing."

During the mission, there was a premonitory breakdown that caused the temperature on board to rise to almost 86°F (30°C), although later the engineers managed to lower it to 73°F (23°C). Getting the temperature under control, their stay of just over two weeks in space was an experience which she enjoyed to the fullest, as she commented in another interview from the space shuttle: "The first couple of days you

---

[1] Jonathan Clark, with his remarkable array of skills, was the medical director of the "Red Bull Stratos" challenge, in which Felix Baumgartner jumped in free fall from a height of 39 km in 2012.

[2] http://www.edition.cnn.com/2003/US/02/01/sprj.colu.profile.clark/index.html

don't always feel too well. I feel wonderful now. The first couple of days you adjust to the fluid shifting, how to fly through space without hitting things or anybody else. But then after a couple of days you get in a groove. It's just an incredibly magical place." Laurel especially enjoyed the views of Earth from space.

The disintegration of Columbia would lead in the long term to the end of the Space Shuttle program. Of the ship, little could be recovered. Astonishingly, some experiments landed intact including several hundred worms that managed to survive the impact. But, alas, the physical constitution of human beings is more delicate, and it was necessary to identify the fragments of the astronaut's bodies that could be recovered by their DNA.

Clark's remains were buried on March 10 of that year, the day of her birthday (she would have turned forty-two), in Arlington National Cemetery[3]. Her husband was part of the commission that investigated the tragedy. Before this he had declared himself a workaholic, but when he became a widower he decided to turn to the care of his only son, Iain, who was then eight years old. A year later, Jonathan would leave NASA.

Laurel, like Kalpana, lived as she wanted. Her family often told her of their concern about taking up a profession as risky as an astronaut, but Laurel had another point of view: "There's a lot of different things that we do during life that could personally harm us and I choose not to stop doing those things. They [referring to her family] have accepted that this is what I want to do."

As was the case with Challenger, in 2006 it was proposed that seven craters on the Moon be baptized with the names of the crew of Columbia. This proposal was accepted that year by the IAU, and seven craters were chosen[4] inside the great Apollo crater, on the far side of the Moon, south of those dedicated to the Challenger crew. Previously, the IAU had labelled asteroids 51823 to 51829 with the name of the Columbia crew; asteroid 51827, discovered in 2001, was named for Laurel Clark. The L. Clark crater, 16 km in diameter, is the furthest south of the Columbia's seven.

[3] Tragedy had already visited the family two years earlier. Laurel's first cousin, Timothy Haviland, with whom she had a close relationship, was killed during the terrorist attacks on the Twin Towers.
[4] Originally, the craters dedicated to the two female astronauts of Columbia were going to be the smallest. However, finally their craters were chosen to be of an intermediate size. Was something changing at last in the IAU mentality?

28

Valentina Vladímirovna Nikolayeva Tereshkova (1937–)

Hey, sky! Take off your hat, I'm coming!
VALENTINA TERESHKOVA, just before taking off in vostok 6 (1963)

When you have been in space, you can see how small and fragile the Earth is. We must not allow this small planet, blue and bright, to be covered by the black ashes of hatred.
Interview with TERESHKOVA (1975)

"Valya" Tereshkova, the only woman of the Moon still alive at the time of this writing (and an exception to the IAU rule stating that craters are named for deceased scientists and polar explorers), and the first woman in space, was born in 1937 in the small rural town of Maslennikovo, near the Russian city of Yaroslavl (then the Soviet Union). Her father, Vladimir Tereshkov, a tractor driver, was mobilized as a soldier during the Second World War and died on the Finnish front when Valentina was only two years old. Her mother, Elena

Figure 112 Valentina Tereshkova. RIA Novosti archive, image #66514/ Alexander Mokletsov/ CC-BY-SA 3.0 [CC BY-SA 3.0 (https://creativecommons.org/ licenses/by-sa/3.0)], via Wikimedia Commons.

Fyodorovna, who worked in a cotton factory, had to take care of her three children, Vladimir, Valentina (the middle child), and Ludmila, under severe economic hardship.

Valentina's childhood was difficult, though that ultimately played to her strengths: she always stood out for being extraordinarily hardworking.

The Women of the Moon. Daniel R. Altschuler Stern and Fernando J. Ballesteros Roselló.
© Daniel R. Altschuler Stern and Fernando J. Ballesteros Roselló 2019. Published in 2019 by Oxford University Press. DOI: 10.1093/oso/9780198844419.001.0001

From when she was very young, she had to help her working mother with housework, so much so that she could not attend school until the age of ten. Lest it seem that school would provide a break from her labor, at this age she also began to sew on demand to help support the family.

At the age of sixteen the small family moved to live in the house of her mother's parents in Yaroslavl; Valentina had to leave school and start work as an apprentice in a tire factory while continuing her studies by correspondence. She only spent a short time working there because the following year, her mother got her a job at a cotton factory where she worked operating a loom. Being a committed communist, Valentina soon became part of the cotton's factory committee of the Communist Youth, and began to advance positions in the party.

In spite of the work at the loom and the responsibilities that she had, Valentina managed to graduate by correspondence from the Technical School of Light Industry. She also took time for her hobbies; in 1959, she joined the Yaroslavl Sports Air Club to learn skydiving, and she made her first jump at the age of twenty-two. She would become an expert parachutist, and her hobby would become the key to her future.

Two years before her first parachute jump, the Soviet Union had put into orbit the first artificial satellite in history, Sputnik 1, launched on October 4, 1957, initiating the space race, a struggle for technological supremacy and international political prestige between the Soviet Union and the United States. The space race would be an extraordinary incentive for technological advance during the next decades. Space had become a new battlefield of the Cold War between the two opposing political–economic ideologies of these nations: capitalism and communism. But on a smaller scale, the space race was a rivalry of creativity between two men: Sergei Korolev (1906–1966) and the controversial Wernher von Braun[1] (1912–1977).

Von Braun is probably the better known of the two. He was the head of the American space program during the space race and the director of the newly founded NASA (from 1960 to 1970). It is less widely publicized that he was also the creator of the infamous Nazi V-2 rockets that bombarded London during the Second World War. Von Braun was a scientist of the Third Reich who was taken to the United States at the end of the Second World War in the framework of the "Paperclip" operation.

---

[1] Michael J. Neufeld (2007). *Von Braun: dreamer of space, engineer of war*. Knopf.

"Paperclip" was the operation by the American military and espionage services to extract Nazi scientists from Germany during the Second World War and its immediate aftermath, and put them in the service of the American side in the emerging Cold War. As a result of the operation, more than 700 Nazi scientists were transferred to the United States, mainly nuclear and rocket scientists (the Soviets were doing the same: between both sides they divided what was left of science in Germany). At the end of the war, Wernher von Braun and the 500 men of his team surrendered to the American army, which, in exchange for their collaboration, would shelve his Nazi past.

Eventually von Braun's team would end up being the brains behind the North American space program that would ultimately land a man on the Moon. We might speculate on whether if Hitler had not lost the war, perhaps a swastika would now be planted on the moon, but it seems unlikely, since (surprisingly) Hitler did not seem to understand the potential of the rockets. Von Braun recalled, decades later, Hitler's reaction to the tests of the rockets that his team developed in Germany: "Normally people were impressed by the noise, but he did not say a word. We showed him a model of the A-3, how it worked and all that, and he listened to my explanation in silence, and at the end he walked away shaking his head. So, I had the distinct impression that he thought something would not work, or that it was a crazy idea or something like that."

Von Braun's opponent, Sergei Korolev, was a great unknown to the West—his very existence was a state secret, for fear that the CIA might kidnap him, and his identity was revealed only after his death. Korolev had been sent in 1938 to a gulag in Siberia during the Stalinist purges, and for six years he lived under brutal conditions that seriously impaired his health for the rest of his life. When Nikita Khrushchev became president of the Soviet Union, the country began its de-Stalinization process and Korolev was not only released from prison, he became the director of the Soviet space program.

Korolev was a true genius who managed to do great things with small means, while at the same time fighting against the Soviet bureaucracy and the terrifying political commissars. Thanks to him, the Soviet side got a fast start in the space race. His first success was Sputnik—the first manmade satellite. This was followed by the orbital flights of the dog Laika in 1957 (the first living being in orbit) and Yuri Gagarin in 1961, the first astronaut in history[2]. After the successful and world-famous

---

[2] Sorry, Cosmonaut.

flight of Gagarin, Nikita Khrushchev (1894–1971) launched his challenge to the world: "And now let the other countries try to catch up with us. Let the capitalist countries try to catch up with our great nation, which has opened the way to outer space"[3].

Korolev did not like to repeat missions. He had the conviction that each space flight should serve to advance the knowledge of *astronautics*, or else it was a waste of time. After the flight of Gagarin, therefore, Korolev began to think about the next step to take in the exploration of space. One of his ideas was sending a woman into space. Such a mission would raise the prestige of the nation, and it would be a clear message to the world that the Soviet Union treated men and women equally, all equal under the light of communism. A female cosmonaut would stand out in stark contrast to NASA's entirely male astronaut body. The Central Committee of the Party saw the publicity advantages of the project and approved Korolev's proposal.

The Soviet people had received with enthusiasm the news that a young Soviet cosmonaut had successfully carried out the first manned space flight in history. Everyone in the cotton factory where Valentina worked took pride and joy in that achievement. Tereshkova herself was stirred by the bravery of the young Gagarin. The night of his flight, when she returned home from work, her mother (who had followed a course of thoughts similar to Korolev's) made an off-the-cuff comment that would change her life forever: "Now that a man has gone into space, the next time a woman must go." Valentina saw her course with crystal clarity: she would volunteer for the Soviet space program.

It was the right time, since Korolev had just launched an initiative to recruit a female team of cosmonauts. Supervised by Yuri Gagarin, the selection process began in mid 1961; since there were not many female pilots, the ranks of Soviet parachutists became an excellent alternative from which to choose candidates[4]. Four hundred candidates applied to the program, and five were selected, the first body of female cosmonauts in history: Tatiana Kuznetsova, Valentina Ponomaryova, Irina Solovyeva,

---

[3] In contrast, for Korolev the space race was the excuse to open space for humanity: "Our scientists have paved the way to space for the first Russian cosmonauts. Someday space flights with passengers and interplanetary communication will be the norm. It will be fascinating to use satellites as relay transmitters and to establish a network, first national and then worldwide, of television and communication. It will be good to put all this at the service of all the nations of the world."

[4] At that time, Valentina had made 120 parachute jumps.

Zhana Yerkina, and Valentina Tereshkova. All of them were experienced amateur parachutists, except for Ponomaryova, who was a civilian pilot.

The candidates, who received the honorary military rank of lieutenant in the Soviet air force, began several months of hard training, including flights in free fall, tests in isolation and in a centrifuge, and training as pilots of jet planes, as well as theoretical classes and practices of astronautics. According to Yuri Gagarin, referring to Tereshkova: "It was hard for her to master the technique of space rockets, to study designs of spacecraft and equipment, but she approached the work with obstinacy, and devoted much of her time to studying, getting involved in the study of books and notes at night."

Finally, the cosmonaut woman who would fly on the next mission had to be selected. It would be a double flight, in which the twin ships Vostok 5 and Vostok 6 would be launched at the same time, to meet (really just a flyby) in space. Initially, the plan was to send two women, but then Korolev pondered that it would be more interesting to make a comparative study of the effects of space flight on the body of men and women, and that it would be better to send a man in Vostok 5 and a woman in the Vostok 6. The cosmonaut chosen for Vostok 5 was Valeri Bikovski (1934–2019).

The woman chosen for the Vostok 6 was, as we already know, Valentina Tereshkova. She was not the brightest of the group, nor the one in the best physical condition; based only on these qualities she would probably have been placed second. But she had a past story that made her perfect as an instrument of publicity for the regime. She was an ex-factory worker, a proletarian of humble origin who had overcome a war-torn family and had trained at the same time as she worked. She had worked her way through the Soviet regime and was also a member of the Communist Youth. Like Gagarin, she was a perfect example of the success a proletarian could achieve under a communist regime[5].

Figure 113 Tereshkova during the Vostok 6 flight.

---

[5] She was, incidentally, ten years younger than the youngest American astronaut, Gordon Cooper.

On the eve of the launch, when Tereshkova left home, she told her mother that she was going to a skydiving competition so that she would not worry. The first craft to take off was Vostok 5. The next day, on Sunday, June 16, 1963, when Bikovski had been in orbit for almost a day, Vostok 6 took off with Valentina, whose call sign (at first, top secret) during the mission was *chaika*, seagull. Once in orbit, Valentina sent her message, which would enter the history books: "Ya Chaika, Ya Chaika [I am Seagull]! I see the horizon [...] This is the Earth; how beautiful it is. Everything goes well." The feat was broadcast and televised throughout the Soviet Union and in several European countries—including the Tereshkova household, where Valentina's mother found out where her daughter really was! In the television images she looked attentive, full of joy, with a notebook and a pencil that floated before her face.

In its first orbit, Vostok 6 approached nearly five kilometers from Vostok 5, the closest point they reached during the mission, and they established radio contact. Tereshkova said that she could even see Bikovski in Vostok 5. Then the ships separated as their orbits evolved and the cosmonauts engaged in biomedical experiments, and learned to live and work in the absence of gravity[6]. Valentina also realized that there was a programming error in the ship's control software, so that when she was ordered to descend, she would actually ascend (oops!). She communicated the issue to Korolev and fortunately the team was able to solve the programming error from the ground[7].

In Korolev's team, Vostok ships were informally (but appropriately) called "tin cans." They were so small that the cosmonaut could only fit in their interior half-lying down, tied to the ejection seat. Despite this, after twenty-four hours in orbit, Valentina Tereshkova passed word to the control of the mission that she felt perfectly well and asked permission to extend the mission for up to three days. This was granted, and she spent the seventy hours and fifty minutes that her mission lasted in this uncomfortable position; she completed a total of forty-eight orbits around our planet. Finally, the reentry and landing sequence began. The computer sent the order to turn on the rockets to slow the capsule

---

[6] One of the results of the mission of the two Vostok ships was to confirm that women had the same resistance as men for the physical and psychological efforts of being in space.

[7] Tereshkova kept this error secret at Korolev's request, for decades, until 2004: "I kept silent, but Eugeni Vasilievich [Krunov] decided to go public, so now I can talk about it."

and begin its descent towards the Earth, while Valentina used manual controls to maintain the entrance angle of the Vostok. After a hard reentry, and once the ship was braked to a reasonable speed, Tereshkova was ejected from the ship and descended to the ground in a parachute.

Upon her return, Tereshkova was cheered in Red Square and received the highest honors for her historic flight. Her feat filled the front pages of newspapers around the world. She was named Hero of the Soviet Union (the highest award of the nation) and was granted the Order of Lenin. As a symbol of Soviet feminism, she went around the world as an ambassador of goodwill, representing gender equality in the Soviet Union, and she received a thunderous ovation at the United Nations. That same year, Tereshkova married cosmonaut Andrián Nikoláyev, and although the match was rumored to be predicated more on propaganda than love, the couple had a daughter, Elena, the first born of two people who had been in space. For this reason, she was of the interest of the doctors of the space program, in case her parents' exposure could have had some effect on her, an interest that vanished when they found that she was completely normal.

For their part, the politicians and engineers of the United States' space program gritted their teeth being at the tail of the space race (well, second). The six Mercury missions that had flown so far (two of them suborbital) had accumulated fifty-three hours of flight together, less than what Tereshkova—a woman!—had done in a single flight of 48 orbits. The six manned Soviet missions accumulated a total of 382 hours. In a memo from the White House, President Kennedy asked Vice President Lyndon Johnson, then chairman of the National Space Council: "Do we have any chance of beating the Soviets by putting a laboratory in space, or giving a turn around the moon, or landing a ship on the moon, or take a rocket to the moon and bring it back, with a man inside? Is there any other space program that promises dramatic results in which we can win?" The key phrase was "dramatic results": Kennedy wanted something that would capture the imagination of the Americans and show the United States clearly beating the Soviets.

The Soviet space program would score yet another point in 1963 with the first spacewalk by cosmonaut Alekséi Leónov, but it soon suffered a major setback: Sergei Korolev died in 1966, aged fifty-nine, as a consequence of the ailments contracted during his years in Siberia. The whole program rested on his shoulders, and with his death, NASA was finally able to gain the upper hand, an advantage that would culminate in the success of the Apollo 11 mission and the landing of the Eagle in the Sea of Tranquility.

As for the female cosmonauts, Korolev's dream was to have a female crew, but, as Tereshkova drily put it, after his death "another person came with other points of view." No women appeared on the flight schedules for the coming years. Valentina complained bitterly about it in an interview in the Russian weekly *Ogonyok*, in 2003: "They forbade me to fly, despite all my protests and complaints. After having once been in space, I desperately wanted to go back there. But it never happened." She did not fly again, nor did any of the other cosmonauts in her group, which disbanded in 1969.

After Tereshkova, it would take nearly twenty years for another woman to go into space. It would be again a Soviet cosmonaut, Svetlana Savitskaya (born 1948), in August 1982. She would not fly alone (for now, Valentina is still the only woman to have made a solo flight into space), but together with two male cosmonauts, Leonidas Popov and Aleksandr Serebrov. This mixed crew caused rivers of ink to run in newspapers around the world, as it was rumored that again the Soviets had been the first to do something in space: to have sex. The Russian government denied it on several occasions. A year later, the first American woman, Sally Ride, would fly on board Challenger, followed by Judith Resnik, whom we have already met.

After her space adventure, Tereshkova worked as a pilot and as a researcher, specializing in the study of stratospheric aerosols and publishing more than fifty scientific articles. At the same time, she got involved in international work to promote women's rights. She was a member of the Committee of Soviet Women, and she became its director in 1977. In 1987, she was named director of the International Union of Culture and Friendship; and in 1991, she became head of the Russian Association of International Cooperation. In 2011, she was elected to the State Duma, the lower house of the Russian Federation, where she continues to work.

June 16, 2013, marked the fiftieth anniversary of the flight of Tereshkova's gull. During a press conference at the United Nations delegation in Vienna, where many scientists and diplomats gathered to pay tribute to her, she said she still had dreams about that space trip. At seventy-six, Tereshkova confessed that she would like to fly to Mars, even if it was only one way: "It's my favorite planet. Most likely, the first flights to Mars are only one way, that's my opinion. I am willing, but unfortunately this will not happen soon. Of course, this is a dream, to travel to

Mars and see if there was life there. And if there was, why is there no longer? What catastrophe took place on that planet?"

The honors received by Valentina Tereshkova have been numerous and an exhaustive list would be quite long. In addition to those already mentioned, we highlight the honorary doctorate from the University of Edinburgh, in 1990, and the position of Olympic torchbearer for the Beijing Games in 2008. For its part, the IAU baptized asteroid 1671 "Chaika" in honor of Valentina's call sign during her flight. As for her crater, as we mentioned in the chapter on nomenclature, at the IAU's 1967 general assembly the Soviet representatives brought forth a lunar map on which they had named most of the features on the hidden face of the Moon after their compatriots. After extensive negotiations with the IAU, an agreement was reached that gave the Russians a somewhat more modest role in the lunar namescape, but one of the names of this lot that survived, due to her nature as a space pioneer, was that of Tereshkova. Her crater is elliptic in shape and quite eroded, it is located on the western margin of Mare Moscoviense as can be seen in Figures 114 and 115.

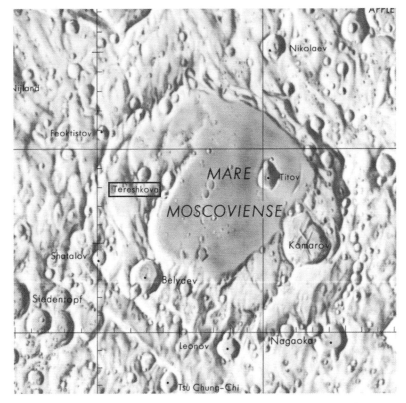

Figure 114 Location of crater Tereshkova. Courtesy of the Lunar and Planetary Institute, Houston, Texas.

Figure 115 Lunar Reconnaissance Orbiter zoom on crater Tereshkova (image width is 150 miles).

Post text

What, if anything, do these women have in common? Actually, few things unite *everyone* in this incongruent set. Of the twenty-eight women of the Moon, two are from the classical era, the Alexandrians Hypatia and St. Catherine, whose craters were named by Riccioli in the sixteenth century (and who, as we have seen, may in historical fact be the same person). The other twenty-six, the modern ones, might better indicate the criteria for deserving a crater on the Moon. Of them, we find seven who are not scientists: five astronauts (except for the pioneer Tereshkova, they were commemorated for their sacrifice, not specifically for their achievements) and two philanthropists of science (we could fairly say they "bought" a crater, without diminishing their contribution to the progress of science). Of the nineteen who were scientists but not astronauts, thirteen were astronomers, and except Lepaute, all of these women were British (four) or American (eight); of the latter, five were from the Harvard Observatory. Of the six who were not astronomers, only Somerville was British and only Cori, American (naturalized). They all were clearly in the first rank of scientists of their generation (two of them won Nobel Prizes and two more indisputably deserved them). The only common factor is the fairly bland observation that all of them had a relation with science (although it seems that being an astronomer of Anglo-American culture makes things easier, and that non-astronomical scientists need a higher level of excellence in their field).

Far more than their commonalities, what stands out from this survey is their astonishingly small number. They are very, very few, and there are a great many women of comparable merit who are not so honored.

One can explain the fact that only twenty-eight of the lunar craters are named after women as a consequence of historical events and social prejudices. But the Moon does not reflect the fact that a significant number of women (despite historical obstacles and social prejudices)

The Women of the Moon. Daniel R. Altschuler Stern and Fernando J. Ballesteros Roselló.
© Daniel R. Altschuler Stern and Fernando J. Ballesteros Roselló 2019. Published in 2019 by Oxford University Press. DOI: 10.1093/oso/9780198844419.001.0001

have contributed to the scientific enterprise. One can look at[1] *Notable Women Scientists*, a book in which we find about five hundred women from around the world, who were selected for their contributions, or the even more extensive[2] *The Biographical Dictionary of Women in Science* with about 2500 entries, to realize there is no shortage of women who have distinguished themselves in science.

The statistics tell us that things did not change with the founding of the IAU. At the time of writing, there were 1586 craters with proper names approved by the IAU. Of these, 567 were approved in 1935, those compiled by Mary Adela Blagg and Karl Müller; that is, they were the names already used prior to the creation of the IAU. In that list of 567 names, only ten of our women appear (1.7% of the total), one of them Blagg herself. After the creation of the IAU and until the present, 1019 new names have been added to that list (of those, around half—515 names—were added in 1970). But of those 1019, only 18 correspond to the women whose lives we have narrated here, which keeps an identical percentage: 1.7% of the total.

It is true that the IAU, in charge of astronomical nomenclature[3], has reserved an entire planet (Venus) on which craters larger than 20 km in diameter honor "deceased women who have made outstanding or fundamental contributions to their field," and craters under 20 km bear "common female first names," while other orographic features (mountains, valleys, plains, etc.) are reserved for mythological goddesses and heroines. The 898 named craters on Venus range from Frances Abington (1737–1815, English artist), through Joliot-Curie, a crater of 91 km in diameter that honors Irène, to Lidiya Zvereva (1890–1916, Russian aviator).

On the other hand, craters on Mars larger than 60 km are reserved for "deceased scientists, especially those who have contributed significantly to the study of Mars; writers and others who have contributed to the lore of Mars." These craters range from Walter S. Adams (1876–1956, American astronomer), through H.G. Wells (1866–1946, English novelist) who we know from *The War of the Worlds*, among other novels, to David D. Wynn Williams (1946–2002, English astrobiologist)[4].

[1] Pamela Proffitt (ed.) (1999). *Notable women scientists*. Gale Group.
[2] Marilyn Ogilvie and Joy Harvey (eds.) (2000). *The biographical dictionary of women in science: pioneering lives from ancient items to the mid-20th century*. Routledge.
[3] Gazetteer of Planetary Nomenclature: http://planetarynames.wr.usgs.gov/.
[4] By the way, on Mars there are only three craters named for women belonging to the twentieth century: Sklodowska for Marie Skłodowska-Curie, Renaudot for a

Other planets and minor bodies of the solar system have their own naming protocols. Thus, on the asteroid Eros, craters are named with "Mythological and legendary names of an erotic nature" (Madame Bovary, Casanova, Lolita, Mahal), and for Pluto, whose amazing surface we have recently imaged with the arrival of NASA's New Horizons, craters will be named for "Scientists and engineers associated with Pluto and the Kuiper Belt" whereas faculae (bright spots), maculae (dark spots), and sulci (furrows and ridges) will be named for "gods, goddesses, and other beings associated with the Underworld from mythology, folklore and literature." And so it goes.

But we feel that it is different—and, yes, better!—to have a crater on the Moon compared to one on another planet, even if it is relatively close to the Earth, like Mars or Venus. The surface of the Moon is accessible to the eye and its craters and other structures are visible with even a modest telescope. We think that a crater of 1 km (approximately the smallest that can be distinguished from the earth with a ten-inch telescope) on the visible face of the moon is preferable to a 100 km crater on another planet. It is estimated that there are over 300,000 lunar craters of that size. There are, therefore, enough to catch up and undo this historical injustice. If, in the words of the first man on the Moon, Neil Armstrong, "We came in peace for all mankind," then all mankind is poorly represented, not only women but all those who live below Earth's equator, even if they are only a small minority. The IAU has a Name Request Form[5] "for use by members of the professional science community" and we wish to encourage others to use it. We have our own favorite candidates who deserve to be distinguished with a crater on the surface of our neighbor. Who is yours?

A note added in proof

The naming of lunar craters (and other planetary features) is an ongoing slow process managed by the IAU, where a Task Group for Lunar Nomenclature currently composed of nine members from several countries reviews proposals for naming. Anyone may suggest that a specific name be considered by the Task Group, but there is no guarantee

French astronomer specializing in Mars, and Sytinskaya for a Soviet astronomer who worked on several robotic probes to Venus and Mars.

[5] https://planetarynames.wr.usgs.gov/FeatureNameRequest

that the name will be accepted. Names successfully reviewed by the Task Group are submitted by the Task Group Chair to the Working Group for Planetary System Nomenclature (WGPSN). Upon successful review by vote of the members of the WGPSN, names are considered approved as official IAU nomenclature, and can be used on maps and in publications.

Since 2014, when this work was originally composed, eleven new lunar craters have been named, and among these, three belong to women (an improvement of sorts). They are named after Marie Tharp (1920–2006), a US geologist and oceanographer who worked at the Lamont–Doherty Earth Observatory (LDEO) of Columbia University, New York, and created the first comprehensive map of the ocean floor and is credited with discovering the Mid-Atlantic Ridge; Elisabetta Pierazzo (1963–2011), an Italian planetary scientist who worked at the Planetary Science Institute in Tucson, Arizona, expert in the area of impact modeling throughout the solar system; and most recently Hildegard von Bingen (1098–1179), a German writer, composer, mystic, and visionary. In Germany, she is considered the founder of the scientific study of natural history. We hope to include their stories in a future edition of this work.

The Moon

Data for the Moon

Orbital elements

Inclination of lunar orbit plane to earth's orbit (the ecliptic)	5.1454°
Eccentricity $e = \sqrt{1 - b^2/a^2}$ (where a is the semi-major axis and b the semi-minor axis of the ellipse). For a circle e=0	0.0549
Orbital synodic period (from full moon to full moon)	29.5 days (so February cannot have blue moons)

Average distance   384,402 km (239,000 mi) [min.= 356,500 km (221,500 mi) max. 406,700 km (252,700 mi)]

Physical characteristics

Mass	7.349×10^{22} kg (0.012 times the Earth)
Density	3.34 g/cm^3 (5.51 for Earth)
Surface area	38 million km^2 (510 million for Earth)
Diameter	3476 km (12,756 for Earth)
Angular diameter (minutes of arc) (for the Sun it is 31.6 to 32.7 and those different numbers account for some solar eclipses being annular and others total)	Perigee (position closest to Earth) 34.1 Apogee (position furthest to Earth) 29.3
Gravity	1.62 m/s^2 (9.8 for Earth)
Escape velocity	2.38 km/s (11.2 for Earth)
Orbital sidereal period (with respect to the stars)	27.3 days
Axial inclination	1.5424°
Albedo	0.12

(Continued)

Orbital elements

Age	4.53 billion years
Composition of the surface (does not contain cheese)	Oxygen 43% Silicon 21% Aluminum 10% Calcium 9% Iron 9% Magnesium 5% Titanium 2%
Atmospheric pressure	3×10^{-10} Pa (note that it is an insignificant atmosphere)
Temperature	When sunlight hits the moon's surface, the temperature can reach 260°F (127°C, above boiling). When the sun goes down, temperatures can dip to −280°F (−173°C)

The Tides

In this bonus appendix, we aim to explain the way that the moon affects life on Earth: the tides. Many people do not understand how the moon affects the seas, and as with any lack of understanding, some will fall into superstition and pseudoscience. One particularly preposterous idea is that since the tides affect the oceans, and humans are mostly water, the lunar tides also affect us in some way, a very bad analogy, and you will learn why.

Let's consider in some detail the effects of the Moon and the Sun on Earth. At these distances and masses, the relevant force is gravity, and the force of gravity between any two objects is, according to Newton's law of gravitation, proportional to each of their masses and the square of the distance between the two objects. In mathematical notation:

$$F = \frac{GMm}{d^2}$$

The gravitational force F between the centers of two masses (M and m) separated by a distance d. (G is a constant that is determined experimentally.) We are used to comparing the gravitational force that Earth exerts on different objects, it is what we familiarly call "weight"—we can use weight and mass almost interchangeably in everyday discourse because in the equation describing gravitational force, the mass M of the Earth, the distance d from the center of the earth[1], and the gravitational constant G all stay the same. The weight on Earth of different objects is proportional to their mass.

What we are interested in is the gravitational force that the Sun and the Moon exert on objects on Earth, and here we have to do some math. You can plug these numbers into your scientific calculator but we've done the calculations as follows. With the knowledge that gravitational constant $G = 6.67384 \times 10^{-11}$ m^3 kg^{-1} s^{-2}, the gravitational force exerted by the Sun (with a mass $M_S = 1.989 \times 10^{30}$ kg, at a distance $d_S = 1.496 \times 10^{11}$ m) on a mass of one kg ($m=1$) on Earth is:

$$F_s = \frac{GM_s}{d_s^2} = 5.92 \text{x} 10^{-3}\,N$$

[1] This is not, at the absolute limits of precision, true: objects weigh ever so slightly less on the top of a tall mountain than they do at sea level. But it's only detectable with very finely calibrated scientific scales.

And the force exerted by the Moon (with mass M_m=7.34767309 × 10²² kg, at a distance d_m=3.844 × 10⁸ m) is:

$$F_m = \frac{GM_m}{d_m^2} = 3.33 \times 10^{-5} \, N$$

where N is the unit of weight that is called Newton (N) (guess why).

The first thing to notice is that the gravitational force of the Sun on an object on Earth is 178 times greater than the force of the Moon (that's why we are in orbit around the Sun and not the Moon). But both of these forces are vanishingly small compared to Earth's gravitation, which is almost 1600 times greater than the gravitational pull of the Sun, and nearly 300,000 times as strong as the pull of the Moon. The forces on a person of 80 kg standing on the Earth's surface due to the Earth, the Sun, and the Moon are 784 N, 0.47 N, and 0.0027 N respectively.

Note that even though we can calculate these forces, we cannot *feel* the force due to the Sun and the smaller one due to the Moon as they do not affect us since we, together with the Earth and the Moon are in free fall (orbit) around the Sun; and we are similarly in free fall around the Moon—it is conventional to think of the smaller object orbiting the larger, but in reality the two objects both orbit around a point (the center of mass or barycenter) that is somewhere between the center of their two masses, though much closer to the center of mass of the larger one.

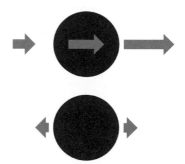

Figure 116 The gravitational pull of the Moon. The top half of the diagram represents the pull of the Moon on objects on the far side of Earth relative to the Moon (left arrow), center of the earth, or on the "sides" of the Earth (center arrow), and the near side of the Earth to the Moon (right arrow). The bottom half of the diagram represents the forces experienced when we subtract the vector representing the pull of the Moon at the center of the Earth.

The next important consideration for us to understand is that the tides are not caused by the gravitational forces that we have previously calculated, but are caused by the *difference* between them on two opposite sides of the Earth, which are separated by some 8000 miles. If that distance were greater, so would be the tides, and it would be smaller if the distance were smaller. The gravitational force of the Moon on the closest side of the Earth is a little larger than the force on the far side (greatly exaggerated in Figure 116). Because (as mentioned) the Earth and the Moon are in free fall around each other, the gravitational force of the moon as experienced at the center point of their orbit (fairly close to the center of the earth) is not experienced anywhere on earth—what this means is that objects on the side of the earth near to the Moon experience only the "extra" gravitational pull by virtue of being closer to the Moon, and objects on the far side of the Moon experience only the "negative" gravity—the centrifugal force in excess of the Moon's gravitational attraction—by virtue of being further away from the Moon. These forces are illustrated in the bottom half of Figure 116.

Although, as we have seen, the gravitational *force* of the Sun is much greater than that of the Moon, the vastly greater distance of the Sun from the Earth means that the tidal effect (which depends on the *difference* in force between two points) of the Moon is approximately twice the tidal effect of the Sun. When there is a full moon (opposition) or a new moon (conjunction), the Sun and the Moon are aligned and the tide is higher because the forces pull in the same direction (spring tide) than when we have a quarter moon, when the forces are perpendicular to each other (neap tide) (Figure 117).

The tidal effect affects mainly the oceans, because fluids deform more easily than solids, and it causes two bulges in the ocean, one on each side of the Earth. As the solid part of the Earth rotates under these bulges, at each fixed point on the surface we observe roughly two high tides a day (due to the movement of the Moon during a day, they occur about every thirteen hours).

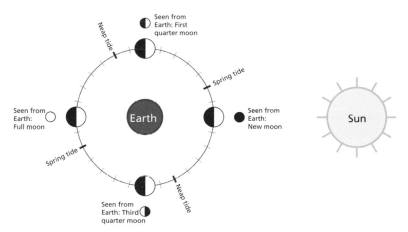

Figure 117 Ideally spring tides should happen exactly at the full and new moons and neap tides are exactly at the one-quarter and three-quarter moons, but due to the bathymetry and frictional effects on coastal areas, neap and spring tides reach their maximum force 2 days after the first quarter moon, third quarter moon and new moon, full moon, respectively.

Note that the Earth itself is also deformed by tidal forces, and the greater these forces are, the greater the deformation. There exists for every mass orbiting another one a distance such that the tidal forces exerted on it are greater than the gravity that holds it together; inside this distance, known as the Roche Limit, the body will disintegrate. Saturn's rings, for example, are within Roche's limit for this planet, and that's why Saturn did not form moons at those distances. In a case where we Earth inhabitants were able to directly observe this process, Comet Shoemaker-Levy 9, which had been captured by Jupiter's gravity and was orbiting the planet, fragmented into several pieces when its orbit brought it too close to Jupiter in 1992; its remains later collided with the planet, in 1994.

The important thing to understand in the case of tidal effect is that it is entirely insignificant at the scale of a human body (a couple of meters compared to 13,000 km) since in this case the difference of the gravitational force between one point and another of the body is vanishingly small. If instead you were falling towards a black hole you would feel a huge difference between head and feet because both of the huge mass many times that of the Sun and because you are very close to it, an effect that has been termed as "spaghettification" or the noodle effect. Your feet fall faster than your head does (or maybe the other way around) but at any rate you will be stretched thin.

Index